T0189350

Transactions on Engineering Technologies

Sio-Iong Ao · Haeng Kon Kim ·
Mahyar A. Amouzegar
Editors

Transactions on Engineering Technologies

World Congress on Engineering and Computer Science 2018

Editors
Sio-Iong Ao
IAENG Secretariat
International Association of Engineers
Hong Kong, Hong Kong

Haeng Kon Kim
School of Software Convergence
Daegu Catholic University
Daegu, Korea (Republic of)

Mahyar A. Amouzegar
Provost
University of New Orleans
New Orleans, LA, USA

ISBN 978-981-15-6850-3 ISBN 978-981-15-6848-0 (eBook)
https://doi.org/10.1007/978-981-15-6848-0

This Springer imprint is published by the registered company Springer Nature Singapore Pte Ltd.
The registered company address is: 152 Beach Road, #21-01/04 Gateway East, Singapore 189721, Singapore

Preface

A large international conference on Advances in Engineering Technologies and Physical Science was held in San Francisco, California, USA, October 23–25, 2018, under the auspices of the World Congress on Engineering and Computer Science (WCECS 2018). The WCECS 2018 is organized by the International Association of Engineers (IAENG). IAENG, originally founded in 1968, is a non-profit international association for the engineers and the computer scientists. The WCECS Congress serves as an excellent platform for the members of the engineering community to meet and exchange ideas. The Congress in its long history has found a right balance between theoretical and application development, which has attracted a diverse group of researchers, leading its rapid expansion. The conference committees have been formed with over two hundred members including research center heads, deans, department heads/chairs, professors, and research scientists from over 30 countries. The full committee list is available at the Congress' Web site: www.iaeng.org/WCECS2018/committee.html. WCECS conference is truly an international meeting with a high level of participation from many countries. The response to the WCECS 2018 conference call for papers was outstanding, with more than four hundred manuscript submissions. All papers went through a rigorous peer-review process, and the overall acceptance rate was 51%.

This volume contains seventeen revised and extended research articles, written by prominent researchers, participating in the congress. Topics include chemical engineering, electrical engineering, communications systems, computer science, engineering mathematics, manufacture engineering, and industrial applications. This book offers the state of the art of tremendous advances in engineering technologies and physical science and applications; it also serves as an exceptional source of reference for researchers and graduate students working with/on engineering technologies and physical science and applications.

Sio-Iong Ao
Haeng Kon Kim
Mahyar A. Amouzegar

Contents

A Novel Multi-layer Detection Scheme for Diffusion-Based Molecular Communications

Mohammed S. Alzaidi(✉), Walid K. M. Ahmed, and Victor B. Lawrence

Stevens Institute of Technology, Hoboken, NJ 07030, USA
{malzaidi,victor.lawrence}@stevens.edu, walidmail@yahoo.com

Abstract. Molecular communication (MC) is the new nano-bio-communication paradigm that inspired by the nature to perform communication tasks at nanoscale networks. Although power utilization and large computation abilities are important factors that play a role in designing detection schemes for traditional communication systems, the MC is more sensitive to these factors due to its operating scale and power resources. Transmitters and receivers can be bio-cellular, bio-organisms, or nano-machines that can perform simple tasks due to their natural limitation with respect to size, power supply, and required simplicity. Accordingly, there is a need for efficient low-complexity detection schemes that can enable reliable communications despite impairments such as inter-symbol interference (ISI), noise and distortion. In this chapter, we propose a novel detection scheme for molecular communication via diffusion (MCvD) that achieves low computational complexity, yet deliver reliable information detection. Our proposed scheme encompasses a multi-layer structure where each layer performs a certain level of pin-pointing (or zoom-in) of the estimated received bits in a fashion that can be regarded as a form of sub-optimal minimum mean-square estimation (MMSE) that tradeoffs complexity for performance. Our detection method has been able to achieve un-coded bit error rate levels down to 10^{-5} at signal-to-noise ratios where traditional threshold-based detection achieves worse than 10^{-2}.

Keywords: Correlation · Diffusion · Entropy metric · Logarithmic metric · Molecular communication · Molecular concentration

1 Introduction

Electromagnetic and acoustic propagations are the primary and centric ways for communication systems. However, their waves still do no propagate effectively due to severe obstruction or path loss [1] for some propagation media such as various types of fluids, biological and tissue-based environments. Nano-molecular communication has been an emerging mechanism for intra-body communications that has been attracting growing attention in the research community [2]. Molecular communication (MC) is based on communicating information in a manner

S.-I. Ao et al. (Eds.): WCECS 2018, *Transactions on Engineering Technologies*, pp. 1–15, 2020.
https://doi.org/10.1007/978-981-15-6848-0_1

inspired by chemical signaling that naturally occurs in our bodies, for example. Information can be conveyed by letting the transmitter (Tx) release a perfect number of molecules of certain types that propagate through mediums such as fluid or gaseous and eventually hit the desired receiver (Rx) that able detect such molecules and determines the information content from the pattern and amount of the received molecules [3–8]. Molecular communications offer a viable link for short, medium and long distances of the order of micro-meters [9]. Moreover, MC is envisioned to be used in a variety of domains including industrial, environmental, and biomedical applications [10–12]. Transmitter (Tx) and receiver (Rx) can be either a biological organism/cell, or fabricated nanomachines (NMs) which can perform simple tasks [13,14] with respect to their tiny resources of power, size, and design complexity [2,15]. Accordingly, any detection scheme that requires a large consumption of power or complex computational operations will not be practical to apply for MC. The design of detection schemes always needs to acquire simplicity and low complexity.

In literature, several detection schemes have been introduced for the diffusion-based channel model. Some of these schemes are new and designed specifically for MC while others have been used previously with the traditional communication systems. Amplitude detection and energy detection schemes have been proposed in [15], the amplitude detection detects each bit by measure the amplitude of the concentration in a pre-specific time instant. The energy detection measures the energy of the received signal of each symbol duration. Both of these detection methods make their decision on bits by comparing signal measurements with a pre-determined threshold. In [18], two coherent detection schemes, the maximum-likelihood (ML) sequence detector and the maximum a posterior detector that commonly used in traditional communications have been applied to the MC. These methods need a large number of computational operations and power consumption, which is not efficient for MC.

In this chapter, we provide the details and evaluate the performance of our novel detection scheme which has been designed based on a multi-layer approach for MC via diffusion. By utilizing a discrimination layer followed by small-size of low-complexity correlation operations layer, our algorithm will be able to perform the lowest operations that are needed to detect received signals. Our performance evaluation results show significant waterfall un-coded bit error rate (BER) down to 10^{-5} at low signal-to-noise ratios where traditional threshold detection achieves worse than 10^{-2}. Moreover, our approach focuses on short signaling periods to increase data rates that most states of art detection schemes in MC avoid it for better performance.

The remainder of the chapter is structured as follows. Section 2 shows the details of the MC system model based on the free diffusion-based channel. In Sect. 3, we introduce a novel multi-layer detection scheme. In Sect. 4, we discuss the simulation results that evaluate our proposed detection scheme while Sect. 5 concludes the chapter.

2 System Model

Our MC system model consists of a single Tx, information molecules, a fluid medium, and single Rx. The Tx is a point source that releases molecules in the medium. The Rx is a NM spherical shape to detect those molecules as illustrated in Fig. 1. We assume that the Tx and the Rx are located in a stationary environment where both are synchronized over time. In this chapter, we apply a binary ON-OFF keying modulation scheme with equally likely transmitted binary information bits. However, we did not apply any coding technique to our system model.

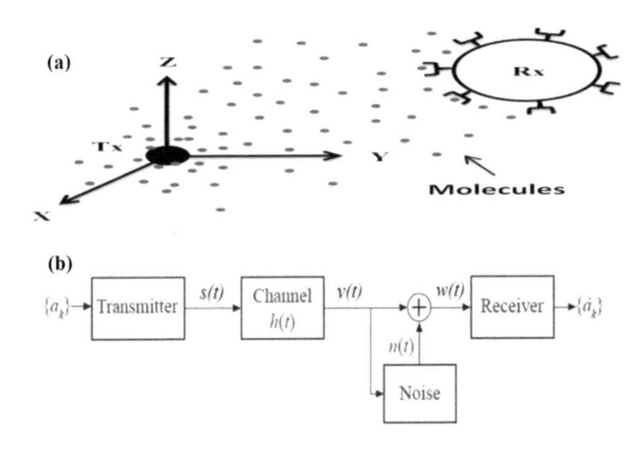

Fig. 1. Diffusion-based MC system components. (a) Graphical illustration. (b) Block diagram of MC system.

The Tx encodes information bits a_k to molecules concentrations pulses, $s(t)$, which can be considered as a rectangular pulse given as [20]:

$$s(t) = Q. \sum_{k=0}^{\infty} a_k.rect(\frac{(t - \frac{T_w}{2})}{T_w - kT_b}) \tag{1}$$

where T_w is the pulse width at the beginning of each symbol duration T_b.

Then, the concentration of the molecule pulses $s(t)$ diffuse randomly based on the Brownian motion process until they hit the Rx. In this type of channel, the propagation relay on some factors such as diffusion coefficient, distance, and time between Tx and Rx. The second Fick's law of diffusion represents the propagation environment [21]. The concentration of the molecules, which is at the location (x,y,z) that diffuse based on the diffusion coefficient D, and time t is given by the equation [22]:

$$h(t; x, y, z) = \frac{Q}{4\pi D t^{\frac{3}{2}}} e^{(\frac{-(x^2+y^2+z^2)}{4Dt})} \tag{2}$$

The response of the diffusion channel $y(t)$ is basically the convolution between the modulated signal $s(t)$ and the impulse response of the diffusion channel $h(t)$. The total received concentration signal that detected by Rx, which is caused by sending original transmitted bits sequence a_k that distorted through the channel is donated by w(t), which can be represented as:

$$w(t) = \sum_{j=0}^{\infty} a_k.y(t - jT_b) + n(t) \tag{3}$$

where $n(t)$ is the counting noise that generated by the random process. It is assumed to be zero-mean Adaptive-white-Gaussian-Noise (AWGN) with a variance of $\sigma_n^2, i.e., n(t)\ N(0, \sigma_n^2)$.

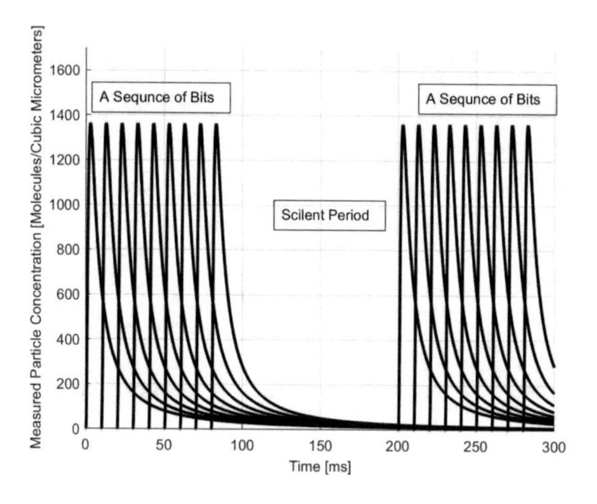

Fig. 2. The impulse response of the diffusion-based channel. The Tx approach sends a number of bits, gets silent for another number of bits, and sends another number of bits again.

To mitigate the inter-symbol interference (ISI), we use the same technique that introduced in [23]. Thus, the Tx sends a number of N-consecutive-bits and then get silent, before sending another N-consecutive-bits to mitigate the ISI as shown in Fig. 2. In this chapter, we call these N-consecutive-bits *words*. The received word samples r_k can be defined as:

$$r_k = y_k + \sum_{i=k-m}^{k-1} y_{k-m} + n_k \tag{4}$$

where y_k is original received bits, $y(k - m)$ is ISI coming from the previous words and n_k is the AWGN noise.

3 The Detection Scheme

Our idea is to detect a sequence of bits together and adapt it to the MC resources limitations. However, detecting a sequence of N-bits will require correlating 2^N possible bit combinations to identify the transmitted bits. Obviously, this approach will consume more power and will require more computational operations which is not the optimal case for the MC. It is worth to mention that the number of possible states will grow rapidly by a factorial factor as N become larger which complicates the detection process. Thus, we have designed our detection scheme by answering two characterization questions about the received bits: How many "bit-1" bits are in the transmitted bits and what are their exact locations in the transmitted sequence of bits? We take "bit-1" in consideration because it is the bit that represents the information molecules as we use the ON-OFF modulation scheme. Each question is answered by passing the received word samples r[k] through two layers respectively that leads to clear detection decisions. The main point in our algorithm is that we used a statistical discrimination metrics, to measure the amount of information in the received word as follows:

1. Entropy metric layer:

$$Entropy_{metric} = \frac{1}{k}(log_{10} \sum_{k=1}^{k} |r_k|) \tag{5}$$

   where k is the number of samples of the received word.
2. Logarithmic metric layer:

$$Logarithmic_{metric} = log_{10}(\frac{1}{k} \sum_{k=1}^{k} |r_k|) \tag{6}$$

   where k is the number of samples of the received word.

In this chapter, we chose the word length case of detection to be eight-bit length to show how our detection method can detect eight bits in a novel and reduced-complexity approach. The eight-bit case will generate two hundred and

Table 1. Zone's patterns based on discrimination values

	Number of "1"s per word	Number of patterns
Zone 1	One	8
Zone 2	Two	28
Zone 3	Three	56
Zone 4	Four	70
Zone 5	Five	56
Zone 6	Six	28
Zone 7	Seven	8
Zone 8	Eight	1

fifty-five bits combinations, excluding the combination of all eight zero bits. Therefore, we calculated the discrimination values for all possible eight-bit combinations by sing both metrics in Eqs. (5) and (6). Then, we noticed that their discrimination values are clustered in eight non-overlapping zones as shown in Table 1. Thus, we can distinguish how many "bit-1" bits occur in each received word by assigning two thresholds to each zone. The minimum and maximum discrimination values of each zone are considered as the lower and the upper thresholds respectively.

3.1 Layer One: Discrimination Layer

The goal of this layer is to distinguish how many bits are in the received word $r[k]$ that equal to "bit-1"? To answer this question, we need to calculate the discrimination value of the received word $r[k]$ by using either the Eq. (5) or the Eq. (6). Then, we can identify the zone that is a candidate to be correlated in the second layer. This process is accomplished by comparing the discrimination value with the zones' threshold values to specify the number "bit-1" bits. The discrimination value of the received word will lead to two possible cases as follow:

- If the discrimination value falls between the upper and the lower threshold values of a specific zone, only the patterns that belong to this zone will be taken to layer two, as indicated in Fig. 3(a).
- If the discrimination value is not related to a specific zone because it falls between two different zones, all patterns that belong to both zones will be merged and taken to the second layer as indicated in Fig. 3(b).

Consequently, the discrimination metric determines a smaller number of bit combinations to correlate instead of correlating all the bits combination.

For the first case, the maximum number of combinations that could possibly be taken to the second layer is thirty-five combinations which are only 13.67% of all bit combinations (86.33% excluded). For the second case, the maximum number of bit combinations that could possibly be taken to the second layer

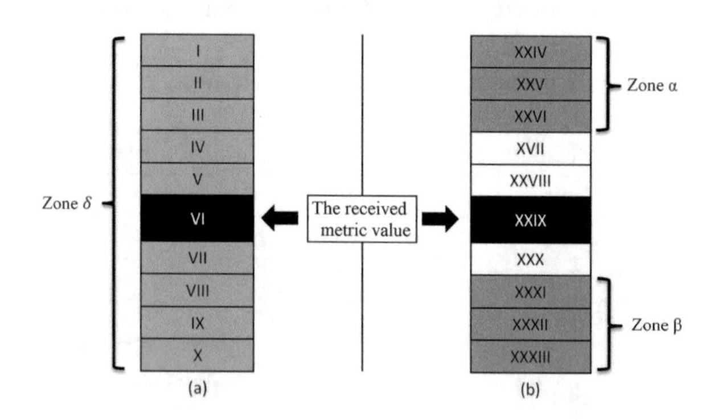

Fig. 3. Scenarios of choosing the zone patterns based on the threshold evaluation

is seventy combinations which are only 27.34% of all bit combinations (72.66% excluded). To point out, our proposed algorithm consumes less power and time to detect a sequence of bits, and there is no need to correlate all bit combinations.

3.2 Layer Two: Correlation Layer

The outcome of layer one classifies the number "bit-1" bits in the received word. Consequently, the goal of Layer two is to identify precisely where the locations of these "bit-1" bits in the sequence. We assume that non-noisy concentration waveforms $z[k]$ for a selected number of bit combinations have been prior stored at the receiver. The correlation calculations in this layer are divided into two stages. The designer of the system has the choice to apply one stage of correlation or two stages based on the case limitations. Every stage of correlation reduces the error probability and help to maintain better performance. The first stage of correlation aims to identify only one candidate pattern from the main bit combinations by calculating the Pearson correlation coefficient between the received word samples $r[k]$ and all main combinations samples $z[k]$ that are related to the assigned candidate zone by layer one as follow:

$$Corr_{coff} = \frac{\sum_{k=1}^{n}(r_k - \bar{r}_k)(z_k - \bar{z}_k)}{\sqrt{\sum_{k=1}^{n}(r_k - \bar{r}_k)^2(z_k - \bar{z}_k)^2}} \tag{7}$$

where $Corr_{coff}$ is the correlation coefficient, $\bar{r_k}$ is the mean value of r_k and $\bar{z_k}$ is the mean value of z_k.

Although our proposed algorithm reduces the number of patterns that are needed for correlation, we still can reduce the number of patterns more if we only deal with the unique patterns. These unique patterns are the main bit combinations that represent more than one combination. For instance, the first zones' combinations are eight bits combinations which all of them have only one "bit-1" digit and seven "bit-0" digits. So, we only store their main bit combination as shown in Table 2.

Table 2. Zone number one patterns

First zone combinations	The unique pattern	Number of patterns
1 0 0 0 0 0 0 0		8
0 1 0 0 0 0 0 0		28
0 0 1 0 0 0 0 0		56
0 0 0 1 0 0 0 0	1 0 0 0 0 0 0 0	70
0 0 0 0 1 0 0 0		56
0 0 0 0 0 1 0 0		28
0 0 0 0 0 0 1 0		8
0 0 0 0 0 0 0 1		1

As shown in Table 3, the difference between the number of all patterns and unique patterns is considered a decent improvement in terms of reducing the number of correlations operations. Additionally, correlating unique patterns instead of all patterns will reduce the probability of errors because of the differences between the patterns will be more prominent in term of the discrimination value and the correlation coefficient.

Table 3. Comparison between the number of all patterns and unique patterns

	Number of "1"s per word	Number of all patterns	Number of unique patterns
Zone 1	One	8	1
Zone 2	Two	28	7
Zone 3	Three	56	21
Zone 4	Four	70	35
Zone 5	Five	56	35
Zone 6	Six	28	21
Zone 7	Seven	8	7
Zone 8	Eight	1	1
		255	128

Thus, using unique patterns will demand us to identify the time lag between the received signal samples and the unique candidate pattern that has the highest correlation coefficient. The second stage of correlation measures the time lag between the received word samples and the candidate pattern samples by using the cross-correlation formula as follow:

$$R_{yz} = \sum_{m=1}^{n} r[m]z[m-L] \tag{8}$$

where L is the lag value between the received signal $r[m]$ and the chosen candidate pattern.

4 Performance Evaluation

This section shows the performance of our novel detection schemes for various parameters that have significant impacts on our proposed method. We used MATLAB as the simulation platform for all evaluation process in addition to using the Monte Carlo simulations approach to get a reliable bit error estimation. All the simulation parameters reflect real parameters for a realistic environment [23]. The distance between Tx and Rx is equal to 3 μm; the diffusion coefficient is equal to 1 $(\text{nm}^2)/\text{ns}$, and the number of released molecules is 1×10^4 molecules.

4.1 The Impact of Symbol Duration Variations

The symbol duration is an essential factor in designing our algorithm; it affects the distances between the zones due to ISI. Thus, we have simulated the BER versus the symbol duration to show the best symbol duration choice that can help to reach the lowest BER. In this manner, we are interested in increasing the high data rate which is a concern in the molecular communication via diffusion (MCvD) due to the ISI. For the Entropy metric, as shown in Fig. 4, we have we evaluated the performance of various symbol durations in the range between 1 ms to 14 ms.

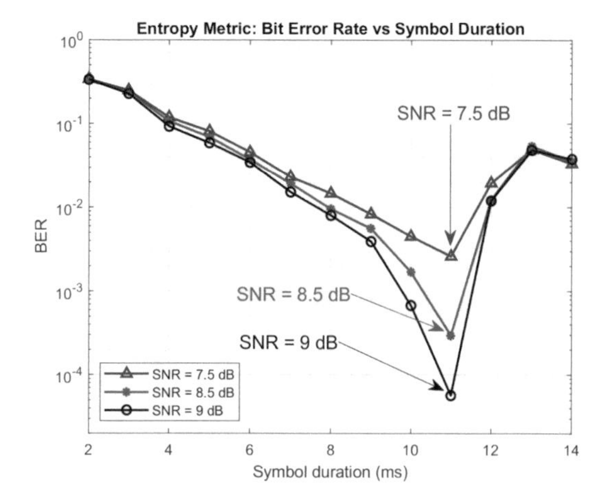

Fig. 4. BER vs. symbol duration for the Entropy metric

As a result, we noticed that the 11 ms is the best symbol duration where the discrimination distances between the zones are the widest, which reduces the BER. Meanwhile, we have simulated the duration between 1 ms and 15 ms for the Logarithmic metric as shown in Fig. 5. Based on the simulation results, we found that at 10 ms, the distances between the zones is the widest, which reduce the BER too.

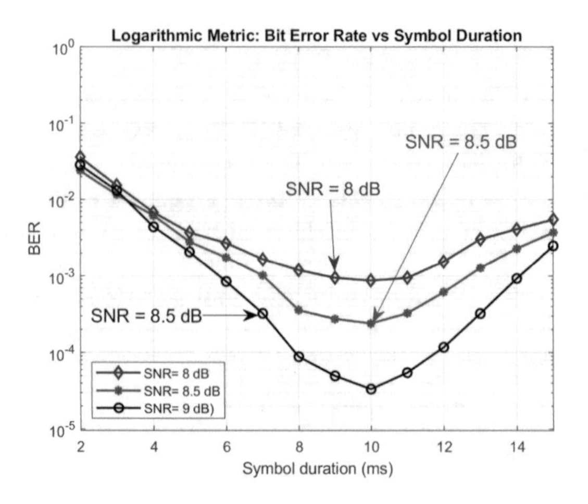

Fig. 5. BER vs. symbol duration for the Logarithmic metric

4.2 The Impact of Signal to Noise Ratio Variations

To show the performance of our scheme in the presence of different levels of the signal to noise ratio (SNR), we have simulated the BER versus SNR for different word lengths such as $N = 6$, 7, 8 and 9. The evaluation performance of the first discrimination metric, Entropy metric, is shown in Fig. 6. It can be clearly seen that the BER of word lengths equal to $6, 7, 8$ is close from each other. In contrast, the BER becomes almost flat at $N = 9$ due to the impact of the ISI on the discrimination distances between the zones.

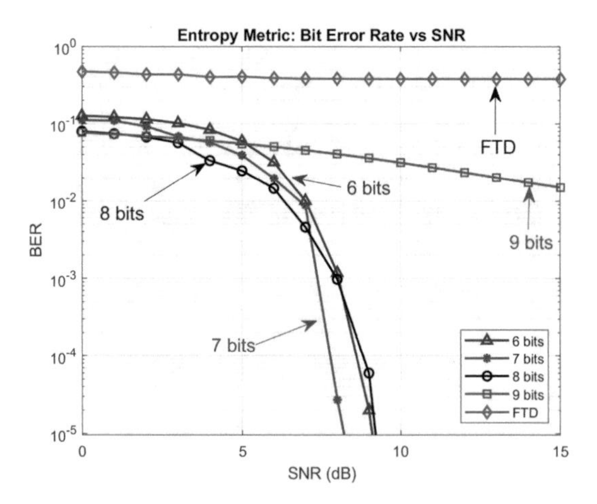

Fig. 6. BER vs. SNR for the Entropy metric

Our proposed algorithm shows a significant low un-coded BER compared to the traditional Fixed-Threshold Detection (FTD) scheme, which has the worst BER. The Logarithmic metric shows a promising performance as shown in Fig. 7 where it reaches lower SNR as 10^{-5} comparing with the Entropy metric. The Logarithmic metric introduces wider discrimination distances between the zones that help to reduce the BER.

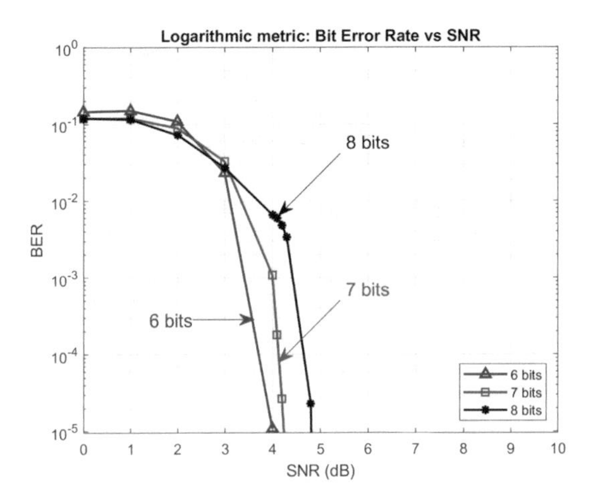

Fig. 7. BER vs. SNR for the Logarithmic metric

4.3 The Impact of Distance Variations

Figure 8 and Fig. 9 show the impact of distance variations for both metrics as they are simulated versus the probability of error. Our proposed scheme shows

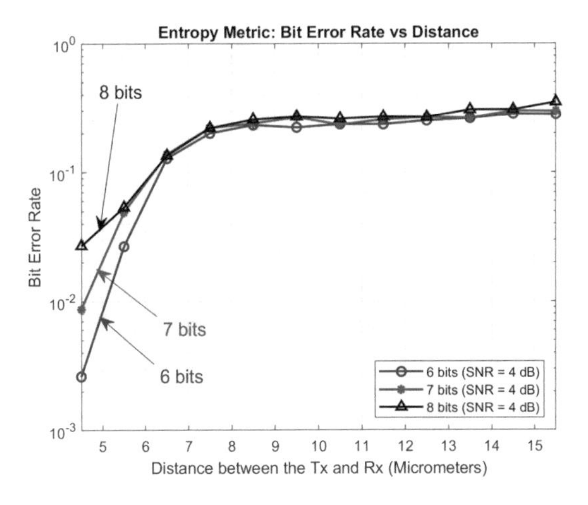

Fig. 8. BER vs. Distance for the Entropy metric

a decent BER as we have evaluated it under a noise to noise ratio equal 4 dB. In both metrics, as the transmitter is getting closer to the receiver, the BER will decrease and vice versa. Our algorithm is designed to perform in short ranges of communications which usually not exceed the size of the cell.

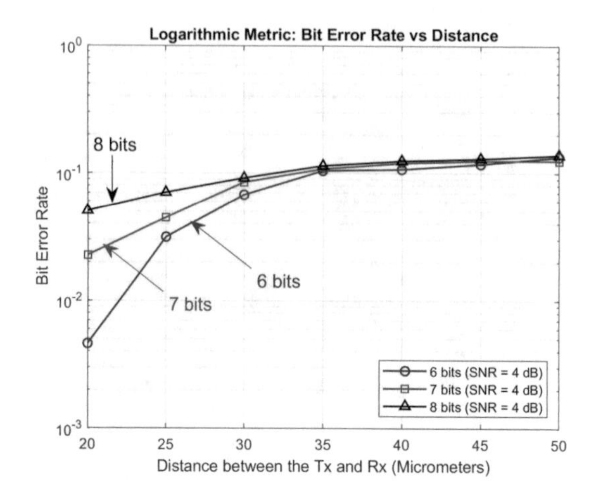

Fig. 9. BER vs. Distance for the Logarithmic metric

5 Conclusion

We proposed a novel detection scheme for molecular communications via the diffusion-based channel. The work aims to develop a reduced complexity fast detection scheme. Based on the entropy metric and Logarithmic metric, the proposed algorithm identifies the required zone to correlate which ultimately identify the transmitted signal. These metrics help to make detection of a group of bits faster by small computational operations in a short measurement period. To date, most groundbreaking studies pursue detection based on a bit-by-bit discrimination approach, which consumes power and time. Our approach can be applied to any modulation scheme and also can be generalized to be applied to any other traditional communication system. Furthermore, our algorithm allows the system developer the ability to choose the number of bits that they would like to detect together (e.g., we simulated N = 6, 7, 8). The simulation results demonstrate that our scheme able to achieve low un-coded BER as 10^{-5} at SNR levels where classic approaches such as FTD scheme achieved BER levels of the order of only 10^{-2} only.

Appendix

This an example for the measurements of a 6-bits length bits that show a clustering in 6 zones by using the Entropy metric.

Table 4. Table for primary patterns of 6-bits Entropy Documentations.

No.	6-bits states						Entropy value
1	1	0	0	0	0	0	0.425031
2	1	1	0	0	0	0	0.540144
3	1	0	0	0	0	1	0.56851
4	1	0	1	0	0	0	0.569986
5	1	0	0	0	1	0	0.580819
6	1	0	0	1	0	0	0.58187
7	1	1	1	0	0	0	0.751513
8	1	0	0	0	1	1	0.770681
9	1	1	0	1	0	0	0.787195
10	1	0	1	1	0	0	0.789276
11	1	0	0	1	1	0	0.794195
12	1	1	0	0	1	0	0.805368
13	1	1	0	0	0	1	0.806243
14	1	0	0	1	0	1	0.817112
15	1	0	1	0	1	0	0.818978
16	1	0	1	0	0	1	0.828056
17	1	1	1	1	0	0	0.930657
18	1	0	0	1	1	1	0.941125
19	1	0	1	1	1	0	0.954485
20	1	1	1	0	1	0	0.958117
21	1	1	0	1	1	0	0.963388
22	1	1	0	0	1	1	0.969308
23	1	1	1	0	0	1	0.970083
24	1	0	1	0	1	1	0.976249
25	1	0	1	1	0	1	0.976761
26	1	1	0	1	0	1	0.984607
27	1	1	1	1	1	0	1.073479
28	1	0	1	1	1	1	1.08204
29	1	1	1	1	0	1	1.094016
30	1	1	0	1	1	1	1.095143
31	1	1	1	0	1	1	1.098036
32	1	1	1	1	1	1	1.189797

References

1. Guo, W., Mias, C., Farsad, N., Wu, J.L.: Molecular versus electromagnetic wave propagation loss in macro-scale environments. IEEE Trans. Mol. Biol. Multi-Scale Commun. **1**(1), 18–25 (2015)

2. Akyildiz, I.F., Brunetti, F., Blázquez, C.: Nanonetworks: a new communication paradigm. Comput. Netw. J. **52**(12), 2260–2279 (2008)
3. Hiyama, S., Moritani, Y., Suda, T., Egashira, R., Enomoto, A., Moore, M., Nakano, T .: Molecular communication. In: Proceeding NSTI Nanotechnology Conference (2005)
4. Nakano, T., Suda, T., Moore, M., Egashira, R., Enomoto, A., Arima, K.: Molecular communication for nanomachines using intercellular calcium signaling. In: Proceeding 5th IEEE Conference on Nanotechnology, pp. 478–481, July 2005
5. Nakano, T., Moore, M., Wei, F., Vasilakos, A., Shuai, J.: Molecular communication and networking: opportunities and challenges. IEEE Trans. NanoBiosci. **11**(2), 135–148 (2012)
6. Nakano, T., Suda, T., Okaie, Y., Moore, M., Vasilakos, A.: Molecular communication among biological nanomachines: a layered architecture and research issues. IEEE Trans. NanoBiosci. **13**(3), 169–197 (2014)
7. Wyatt, T.D.: Fifty years of pheromones. Nature **457**(7227), 262–263 (2009)
8. Nakano, T., Eckford, A.W., Haraguchi, T.: Molecular Communication. Cambridge University Press, Cambridge, UK (2013)
9. Veiseh, O., Gunn, J.W., Zhang, M.: Design and fabrication of magnetic nanoparticles for targeted drug delivery and imaging. Adv. Drug Deliv. Rev. **62**(3), 284–304 (2010)
10. Wang, J., Yin, B., Peng, M.: Diffusion based molecular communication: principle, key technologies, and challenges, China Commun. **14**(2), 1–18 (2017)
11. Yager, P., et al.: Microfluidic diagnostic technologies for global public health. Nature **442**(7101), 412–418 (2006)
12. Farsad, N., Yilmaz, H.B., Eckford, A., Chae, C.B., Guo, W.: A comprehensive survey of recent advancements in molecular communication. IEEE Commun. Surv. Tutor. **18**(3), 1887–1919 (2016). Third Quater
13. Suda, T., Moore, M., Nakano, T., Egashira, R., Enomoto, A., Hiyama, S., Moritani, Y.: Exploratory research on molecular communication between nanomachines. In: Proceeding Genetic and Evolutionary Computation Conference (GECCO), Late Breaking Papers, vol. 25, p. 29 (2005)
14. Nakano, T., Moore, M., Enomoto, A., Suda, T.: Molecular Communication Technology as a Biological ICT. Springer, New York (2011)
15. Llatser, I., Cabellos-Aparicio, A., Pierobon, M., Alarcon, E.: Detection techniques for diffusion-based molecular communication. IEEE J. Sel. Areas Commun. **31**(12), 726–734 (2013)
16. Zheng, O.Z., Ali, M., Basu, K.: Comparing the complexity of two network architectures. Ann. Emerg. Technol. Comput. (AETiC) **1**, 7–18 (2017)
17. Kuzmics, G., Ali, M.: Intra-building people localisation using personal bluetooth low energy (BLE) devices. Ann. Emerg. Technol. Computi. (AETiC) **2**, pp. 24-36, April 1 2018
18. Kilinc, D., Akan, O.: Receiver design for molecular communication. IEEE J. Sel. Areas Commun. **31**(12), 705–714 (2013)
19. Guo, W., Asyhari, T., Farsad, N., Yilmaz, H.B., Li, B., Eckford, A., Chae, C.B.: Molecular communications: channel model and physical layer techniques. IEEE Wirel. Commun. **23**(4), 120–127 (2016)
20. Pierobon, M., Akyildiz, I.F.: A physical end-to-end model for molecular communication in nanonetworks. IEEE J. Sel. Areas Commun. **28**, 602–611 (2010)
21. Bossert, W.H., Wilson, E.O.: The analysis of olfactory communication among animals. J. Theor. Biol. **5**(3), 443–69 (1963)

22. Mahfuz, M.U.: Achievable strength-based signal detection in quantity-constrained PAM OOK concentration-encoded molecular communication. IEEE Trans. Nanobiosci. **15**(7), 619–626 (2016)
23. He, P., et al.: Improving reliability performance of diffusion based molecular communication with adaptive threshold variation algorithm. Int. J. Commun. Syst. **29**(18), 2669–2680 (2016)
24. Alzaidi, M.S., Ahmed, W.K.M., Lawrence, V.B.: Achieving very low un-coded BER via a novel reduced-complexity fast-detection for diffusion-based molecular communications. In: Proceedings of The World Congress on Engineering and Computer Science 2018, Lecture Notes in Engineering and Computer Science. San Francisco, USA, 23-25 October pp. 24-29 2018

Enterprise Mobile Ad-Hoc Implementation

William R. Simpson(✉) and Kevin E. Foltz

Institute for Defense Analyses, 4850 Mark Center Drive, Alexandria, VA 22311, USA
{rsimpson,kfoltz}@ida.org

Abstract. Threat intrusions have led to a formulation of guarded enterprise systems. The approach was to build an impenetrable fortress to prevent hostile entities from entering the enterprise domain. However, this defense and its many re-enforcements have repeatedly been found to be inadequate. The current complexity level has made the fortress approach to security, which is implemented throughout the defense, banking, and other high-trust industries unworkable. An alternative security approach, called Enterprise Level Security (ELS), is the result of a concerted multi-year program of pilots and research. The primary identity credential for ELS is the Public Key Infrastructure (PKI) certificate, issued to the individual who is provided with a Personal Identity Verification (PIV) card with a hardware chip for storing the private key. All sessions are preceded by a PKI mutual authentication (secondary authentication may be employed when necessary), within Transport Layer Security (TLS) 1.2, and a secure communication pipeline is established. This process was deemed to provide a high enough identity assurance to proceed. However, mobile ad hoc networking allows entities to dynamically connect and reconfigure connections to make use of available networking resources in a changing environment. These networks range from tiny sensors setting up communications based on a random or unknown configuration to aircraft communicating with each other, the ground, and satellites. Scenarios have differing requirements in terms of setup, reconfiguration, power, speed, and range. This paper presents an adaptation of the ELS principles to the mobile ad hoc scenario.

Keywords: Enterprise Level Security · Field connectivity · Mobile Ad-hoc · Mobil nexus · Networking · Service requirements

1 Introduction

Mobile ad hoc networking includes a broad range of possible implementations. These implementations range from unstructured networks like MANETs [1], where there is no existing infrastructure and nodes must dynamically configure themselves into a functioning network, to situations in which a mobile node connects to existing infrastructure. This document focuses on situations in which nodes come in and out of communication range of fixed infrastructure and situations in which nodes dynamically connect and disconnect to each other and different networks. These situations allow many of the higher-layer functional and security protocols to function properly. The following

S.-I. Ao et al. (Eds.): WCECS 2018, *Transactions on Engineering Technologies*, pp. 16–29, 2020.
https://doi.org/10.1007/978-981-15-6848-0_2

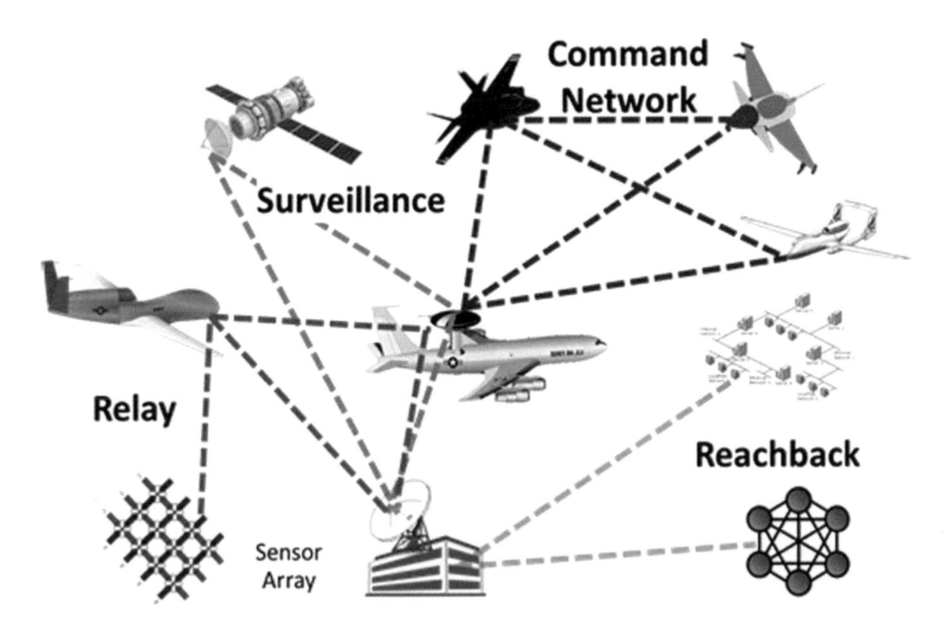

Fig. 1. Ad-Hoc networking

sections describe different aspects of the networking infrastructure that together support the concept of ad hoc connections and mobility. Figure 1 illustrates those network types.

Enterprise Level Security (ELS) is a capability designed to counter adversarial threats by protecting applications and data with a dynamic claims-based access control (CBAC) solution. ELS helps provide a high assurance environment in which information can be generated, exchanged, processed, and used. It is important to note that the ELS design is based on a set of high-level tenets that are the overarching guidance for every decision made, from protocol selection to product configuration and use [2].

This chapter is based in part on a paper published by WCECS 2018 [3].

2 Mobile Ad Hoc Network Services

The services described in this section are shown in Fig. 3. These services are automated, and seek operator confirmations only when and if required. They reside on each element participating in the networks shown in Fig. 2. Each element in Fig. 3 must participate in a handshake with the nexus (see Sect. 2.1) that identifies compatible protocols, waveforms, and drivers to establish a connection. These services act as the initial end-points for connection management. The connection is followed by a bi-lateral authentication and secure channel to the end-point device manager service [4]. The end-point device manager service is the entry point for the requester to access domain services. This must be followed by bi-lateral authentication at the device level. Basic services are shown on the left, building from basic hardware capabilities to supported protocols. Mobile

Ad Hoc Network services are on the right, building from hardware and software management to the "Send Data" service that takes data and a destination as an input and sets up appropriate connections and initiates the communication using the supplied data. Arrows indicate dependencies, where arrows point from the service that is used to the service that uses it.

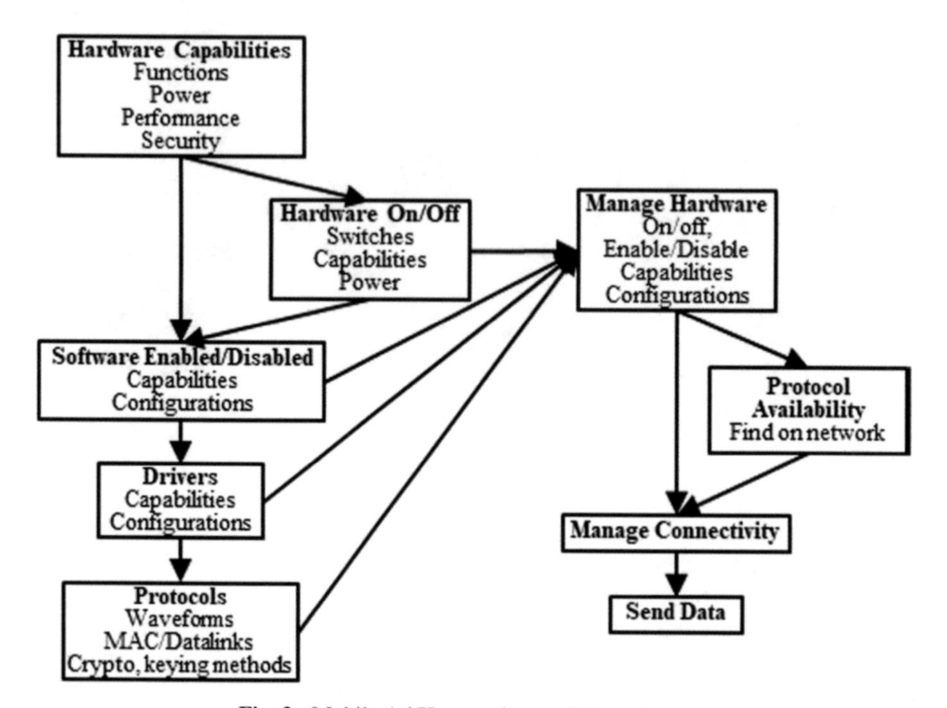

Fig. 2. Mobile Ad Hoc services and dependencies

2.1 The Nexus Requirement

Certain members of the networks are designated as nexus. Nexus points may be located throughout the operational area, and they seek out and provide handshakes to any other nexus points within range. The chaining of nexus points allows reach back from the local network to the enterprise when one or more of the nexus in the chain can reach a network node. The member must have full system capability and acts as the manager of ad-hoc sub networks. An end-point device manager service [5] must reside on a nexus, and a nexus must be part of each network and is the entry point for the requester to access domain capabilities. An example of designated nexus points is shown in Fig. 3.

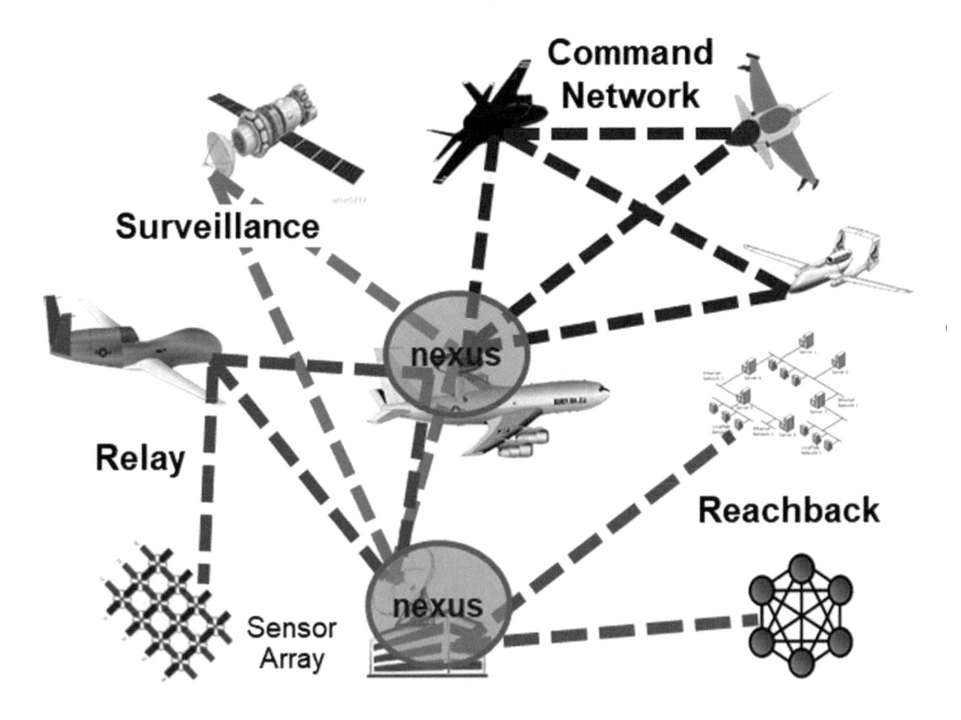

Fig. 3. Designated nexus members.

2.2 Detection of Hardware Capabilities

This section describes the basic services that provide information about the available hardware and the software that directly controls it. These are typically duplicated for each piece of communication hardware in a device so that higher layer services have direct and independent control over each hardware interface. The interfaces to the hardware may be specific to the hardware so that higher layer services provide and use mediation services to interface with these lower-level services.

In order for a node to join a mobile or ad hoc network it must know that the network exists. This can be done by continuously polling for available connections or looking for connections when a request is made to connect. Polling involves more ongoing work and power but provides continuous feedback, while on-demand connection uses fewer resources but requires explicit instruction and incurs a delay. To bridge these two methods, a local service can be invoked that periodically polls for connections and provides the latest data to higher-layer services. This provides a configurable method to tradeoff between power and responsiveness across all possible connection types.

Connections at the lowest layer involve the hardware that actually does the signal generation and transmission. This hardware is controlled by drivers or other software that provide an interface to the operating system and local applications and services. The following information is of interest:

- Hardware capabilities that exist for a given device:
  - Capabilities that are supported,
  - Power and other performance that is supported;
- Hardware that is enabled or disabled by physical switches or other hardware mechanisms;
- Hardware that is enabled or disabled by software:
  - Capabilities that are enabled or disabled in the software;
- Hardware that has the appropriate software drivers and other code in place for use:
  - Capabilities that are supported by the drivers or software;
- Protocols that are supported for the hardware:
  - Waveforms,
  - Mandatory Access Control (MAC)/Datalink protocols and versions,
  - Crypto protocols, versions, keying methods.

All of these must be exchanged between Ad hoc participants and the nexus points. The nexus acts as the controller for sub network communications. Each of these translates into local services for mobile and ad hoc networking. The services described in this section provide basic information about what networking is available, what could be made available, and the capabilities associated with what is and what could be available. In addition, some configuration of the lower layer hardware and software is made available through these services to other services.

The capabilities list for a device describes what hardware is available. This may take different forms. For some devices it could provide a list of standard hardware regardless of what is currently available, such as standard-issue mass-produced units. Such a service would rely on outside or fixed data sets and not the system itself. Other services describe the hardware interfaces associated with the device. For example, a description of whether Universal Serial Bus (USB) 3.0 is supported or just USB 2.0 would be useful when deciding which hardware device to attach through a USB port. Such services could be offline, static, or based on querying the actual device to determine what is available. Other services describe what hardware is actually connected. Unlike some of the services described above that rely on fixed or external information sources, this service actually queries the system to determine what is connected. For hardware that is found, additional information can sometimes be provided, such as the capabilities of the hardware in terms of speed, power, or supported frequencies.

In some cases hardware is available but switched off. This service provides information about the current state of such hardware. In some cases, hardware that is switched off is indistinguishable from hardware that is not present, but when possible, a distinction is made. This allows a service to inform a user that a physical action must be taken to enable communication.

In addition to hardware switches, there are ways to enable and disable communication hardware through the use of software. This can be through an application, the operating system, registry items, or device driver settings. A service is provided to describe the current state of the communication hardware and allow changing this state as permitted through software. In addition to a simple on/off switch, software can provide detailed capability and configuration information, such as frequencies, versions, protocols, security settings, and many others.

In order to use the communication hardware, appropriate drivers and other software must be available and correctly functioning. This service checks hardware for proper operation and reports the status of the hardware and its drivers. This service may simply examine the driver and perform what amounts to static analysis of the system, or it may actually attempt to use the system and check that it responds appropriately. This service provides not just information about the system, but information about how it is currently operating. This includes whether the device is functioning, as well as which of its capabilities are working, such as transmission speeds, error rates, or power consumption, and potentially how well they are working.

This service provides information about particular protocols that run over different communication hardware. The protocols of interest are the protocols specific to the communication hardware. For example, a Wi-Fi protocol service would provide information about the Wi-Fi protocol, not Internet Protocol (IP) or Transmission Control Protocol (TCP). This service provides information about which protocols are supported by the hardware and which versions of each of these supported protocols are available. Additional information includes which frequencies, waveforms, data link, or MAC layer protocols are supported, and what type of cryptography or other cryptographic protections are available.

2.3 Detection of Network Opportunities

This service provides the ability to test enabled hardware for its protocol support at the network layer. This goes beyond the protocol-based services discussed in the previous section, which apply to the hardware protocols. It looks, for example, for Dynamic Host Configuration Protocol (DHCP) servers, network gateways, DNS servers, and other services that would be available in the presence of a network. These are the services that will be used for web service and web application requests. It is important to know whether these services are available, and to what extent they are provided. Knowledge about whether the connection is local or connected to other networks provides important information about the type of connection that can be used by other services.

This service includes tests for proxies, gateways, and other forms of network intermediaries. For example, proxies can be detected by accessing known sites and checking the certificate provided through Hypertext Transfer Protocol Secure (HTTPS). If it does not match the known good certificate, then a proxy is in the middle. This informs decisions about which network to use, since networks with proxies make ELS communication impossible by preventing end-to-end authentication through TLS, but they would be acceptable for low-security traffic.

2.4 Selection of Waveforms and Protocols

This service is used to turn hardware on and off in order to use a specific set of communication hardware. In some cases this capability can function fully in software using the software interfaces described in the previous section. In other cases, in which physical action is required, a notification to a human or other interface, such as a machine or robot, is required to initiate the hardware action. In either case, the goal is to have the appropriate hardware on and enabled and everything else off or disabled. This can be for power conservation, stealth, or just a general security practice to reduce unneeded interfaces.

In addition to just turning hardware on and off, this service allows configuration of the hardware, to include selection of frequencies, protocol versions, waveforms, and other hardware-level information. This service acts somewhat like a mediation service that provides a standard interface for higher-level protocols to manage the underlying hardware. It translates the hardware and low-level software controls into standard interfaces for the higher layers. This enables a consistent treatment of communication channels and re-use of higher-layer services across the enterprise and different devices within it. This service dynamically maintains a set of connections that provide an optimal allocation of resources to available potential connections based on provided performance metrics. For example, if high-speed connectivity to a particular IP address is desired, the service may continuously poll for available connections and choose the fastest one that has connectivity to the desired endpoint. Other parameters can be weighed against each other as well, such as power consumption, cost, and combinations such as power per bit or power per bit/sec. Additional inputs would be required for this service to operate effectively, including power consumption models, pricing models, and latency and throughput measurements and models.

This service uses the Manage Hardware service to actually make changes to the system and its connectivity. It uses a set of defined metrics, measured and provided information about the available networks and connections, and optimization logic to make decisions about how to invoke Manage Hardware to best provide what is desired.

This service not only determines which protocols are available, as described above, but also performs handshakes and information exchanges to establish IP addresses, secure connections, and other functions that actually initiate protocols for connectivity. Examples include Dynamic Host Configuration Protocol (DHCP) requests, Domain Name System (DNS) queries, and other protocols that are common first steps toward data transfer after initial basic connectivity is established. Any ongoing "ping"-type communication is handled by this service as well to establish and update what protocols are available.

2.5 Service Discovery

Lower-level service discovery is addressed by the Protocol Availability service, but for ELS web services a separate method must be used. In a connected network the claims query service is used to determine a list of all applications and services to which an ELS requester entity has claims or access through identity. In a Disconnected, Intermittent or Limited bandwidth (DIL) mobile ad hoc environment, this service may not be accessible, but a local copy may be available. If so, this can be used for service discovery. This local copy must be hosted in a canonical place that is accessible to anyone on the network so that it can be used as an initial access point to any other ELS services and applications available in the local environment. Although the claims query service is not part of Mobile Ad Hoc services (it is part of the ELS suite of services), it is mentioned here for context. For all communication, the Send Data service is used to choose the hardware, protocol, and associated settings to provide the data transmission and receiving of any associated responses.

This service provides network communication based on any request and uses available connections to send and receive data. Software on a device calls this service to perform any network-based communication, and this service handles all network requests, sets up appropriate connections if available, and takes care of sending the requests and receiving the responses. It notifies the end entity making requests of the status of the current connections. It uses the metrics and parameters for performance, cost, and power as input and passes these on to the Manage Connectivity service to allow it to maintain a set of appropriate connections for communication. However, the Send Data service can override these settings based on current requests. For example, if cost and power are a primary concern, most communications will be disabled by Manage Connectivity. However, when a short high-priority message must be sent on a hardware module that is disabled, Send Data can override the default settings and make performance for that communication a priority for the duration required for the communication.

2.6 Query/Response Capabilities

Like the service discovery described above, query and response capabilities are based on ELS. After mobile ad hoc services are used to establish connectivity ELS queries can proceed. If network connectivity provides access to the Enterprise Attribute Store (EAS) and other network resources, then a standard ELS query can follow. If the local network is isolated and has its own EAS instance, then the local instance can be used to provide ELS-based access to local resources. If the local network is isolated and does not host its own instance of EAS, then access is limited to the non-ELS services provided on the local network. For intermittent connectivity, asynchronous messaging may be offered as a service even if synchronous communication is not, since asynchronous communication can be queued until connectivity returns. As with service discovery, the Send Data service handles the sending and receiving of data over the appropriate connections. The following sections describe the steps in setting up a connection. It is expected that this service will handle all of these either directly or indirectly using the previously mentioned services.

2.7 Network Broadcast

The first step for a mobile or ad hoc connection is for the network to identify itself to the mobile node. This is typically done through some sort of network broadcast that identifies the transmitter, the network it represents, its address, the protocols supported, the security offered and required, and other relevant information. For Wi-Fi, for example, a beacon message is sent 100 times per second with this type of information. In some cases, this function is disabled or limited. For Wi-Fi, the Service Set Identifier (SSID) can be hidden so that only nodes that explicitly request the proper ID are allowed to connect. The beacons can be disabled entirely so that the mobile node must know of the network's existence in advance in order to connect. Other techniques exist to either hide connections or make detection and connection more difficult for unauthorized entities.

These are more difficult to implement on wireless networks because the communications are broadcast to an entity in the vicinity, making replay attacks possible. In general, security protocols are a more robust method of limiting access than simple message content-, formatting-, or timing-based methods. Wi-Fi Protected Access (WPA) for Wi-Fi and IP Security (IPSec) for IP-based network layer communications are examples of such security protocols. For wired networks, security is often minimal, allowing anyone with physical access and connectivity to use available network services. An Ethernet connection usually is initiated automatically when a wire is plugged in to an Ethernet port. Higher-layer services may require further actions for access, but the lower-level connectivity provides little, if any, security.

2.8 System Discovery

After the network identifies itself, if it chooses to do so, the mobile node must discover what is available and how to connect [6–9]. With current systems, many possible network connections are available, such as Satellite, Wi-Fi, Military Link Systems, Broadband, and others. The networks provide information about different connections, and the node must make sense of this and discover which networks are accessible, which protocols and options are supported, which security is supported and sufficient to meet policy requirements, and which connections support higher-layer applications. ELS requires bi-lateral authentication, but it may be based on Identity for access.

2.9 Request to Join

The mobile node, though some internal logic, determines which network to join and initiates a "request to join" handshake [9, 14]. This may involve the exchange of identification information, it may include security parameter negotiation, and it may include protocol negotiation. Wi-Fi often includes security information. Link systems use device profiles to set the message formats and protocols. In any case, this is where the connection from the mobile node to the network node is established, along with any required parameters.

As part of the request to join, physical layer attributes may be collected, such as signal strength, noise level, signal quality, multi-path parameters, location information, and supported waveforms and formats. Wi-Fi 802.11n and 802.11ai support beamforming, allowing the multiple antennas at the transmitter and receiver to be used to determine the direction of transmission, which can boost the signal in the vicinity of the communicating entities and reduce it elsewhere. This allows reduced power, slightly increased security, and potentially better use of available network resources by reducing interference with other transmissions.

Other more advanced techniques may allow the use of multipath and complicated urban obstacles to be used to enhance channel security, quality, power efficiency, and data rates. The transmitter sends a test signal to the receiver, which then relays the received signal properties back to the transmitter. The transmitter can then reshape the transmission to "invert" the environmental distortion and allow positive reconstruction of signals at the receiver. Listeners at other physical locations will not be able to properly reconstruct the signal. This allows lower power transmission, better signal to noise, and potentially better privacy against eavesdroppers.

3 Certificate Exchange

One important part of the request to join includes the exchange of certificates. The certificates are assigned to devices and allow authentication based on a trusted certificate authority. For ELS, certificates are stored in hardware, such as a Hardware Security Module (HSM) [10] or a PIV card [11]. For lower layer exchanges, the device Trusted Platform Module (TPM) [12] is the preferred location. Each device is equipped with a TPM or TPM-like hardware certificate and key store, which is used to authenticate to the network or to the mobile node when required.

For mobile devices without hardware stores a derived credential may be used for the certificate exchange. This derived credential is issued by a trusted registration authority (RA) in the enterprise. The derived credential uses the same original certification as the primary credential. If the primary credential is revoked for reasons relating to certification, the derived credential is also revoked, since its certification is no longer secure. If the primary credential is revoked due to issues specific to the credential instance, then the derived credential may remain valid independently. Revocation of the derived credential similarly may or may not lead to revocation of the primary credential, based on the reasons for revocation.

4 Device Requirements

Devices allowed to join enterprise networks are registered and managed by the enterprise use restrictions. All devices have a PKI certificate (Issued PKI or derived) in hardware storage (preferably in a TPM). The device and the domain controller perform bi-lateral PKI-based Mutual authentication before establishment of the channel to the end-point device manager service. The device may also contain one or more individual user certificates (Issued PKI or derived) that are activated when the user signs on to the device. The device may be required to register with the enterprise domain and report attestation from the TPM and other data such as location (where appropriate).

After joining the network and properly authenticating, it may be desirable to set up an end-point device manager service connection to a remote network. This provides an IP-layer secure tunnel through which higher layer data can be sent. The initial network connection only applies to the link layer, or device-to-device connection.

The end-point device manager connection uses machine certificates to authenticate the mobile node to the end-point device manager server and the end-point device manager server to the mobile node. The end-point device manager server then makes internal network services available to the mobile node. Particular attention must be paid to which nodes are allowed to connect to the end-point device manager server. The devices must have controls, through mobile device management or some other verifiable machine hardware and software integrity checks that ensure that the device is protected from compromise to a level comparable to that of the internal nodes on the network.

5 Discovery of Services

After connecting through the end-point device manager, or just to the local network, service discovery can begin. This starts the use of higher-layer protocols, which talk using various protocols over TCP, UDP, or other transport layer protocols. All active entities must have a credential (derived credentials for entities residing on mobile platforms are permitted) to initiate a request. For example, the requester may use a known URL, such as the EAS Claims Query service to retrieve a list of available services. These services are provided based on the requesting entity's identity, as provided in a PIV, a Non-Person Entity (NPE) certificate, or derived credential, HSM, or other certificate or key store.

Service discovery [13–16] can be initiated locally for DIL environments with a local cache of the claims repository and EAS Claims Query service. The claims query service may be modified to provide identity-based access-only claims. For mobile devices that are provided network connectivity to the primary EAS instance, no cache is required and a normal request is sent. Discovery may be accomplished initially using a Claims Query service. The initial handshake is bi-lateral PKI mutual authentication. This service is identity-based and returns links to claims for service that the requester has. The requester must know the local Uniform Resource Locator (URL) for that service in the connected network.

6 Service Request

When access to the EAS is established, the request for service can be sent to the desired application or service or a link in the Claims Query Service return page may be executed. The EAS-provided link redirects to a Security Token server (STS), which provides authorization information in a Security Assertion Markup Language (SAML), and then redirects back to the service. The service's ELS handler processes the request and allows access.

Mobile and ad hoc networking requires some level of performance to support higher-layer protocols and applications [14, 15]. In some cases, such as poor wireless links or intermittent connectivity, the networking protocols do not function well enough to support the higher-layer protocols. In other cases, the implementation of the protocols is inefficient, uses improper configuration, or adds extra components that reduce performance, such as monitoring or filtering. Those factors under the control of the implementer must combine with those not under control to provide a level of service that supports higher-level protocols and applications appropriate for the network and network participants (Fig. 4).

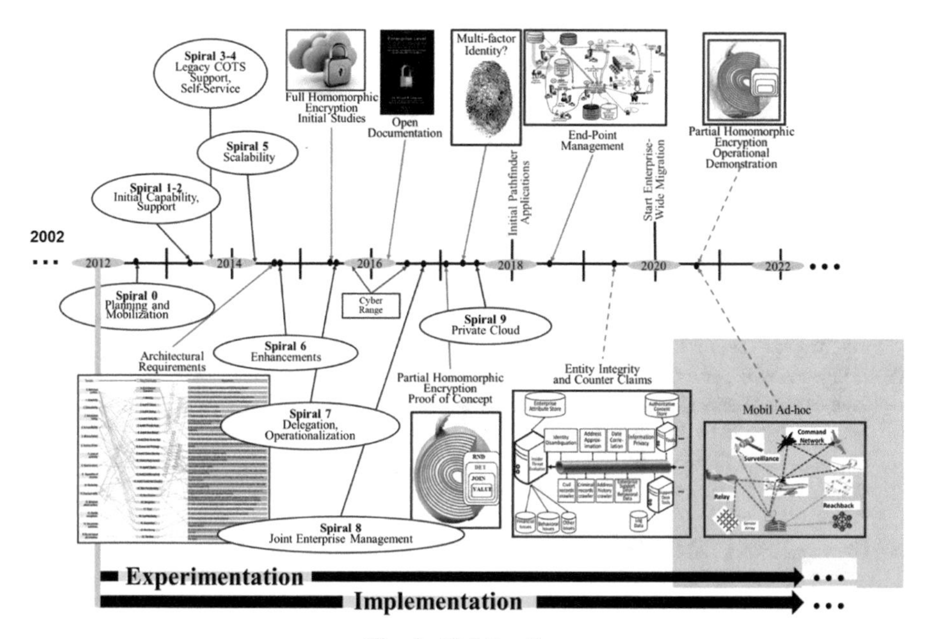

Fig. 4. ELS timeline.

7 Conclusions

We have reviewed the mobile ad hoc issues in a high assurance security system. We have also described an approach that relies on high-assurance architectures and the protection elements they provide through PKI. The basic approach becomes compromised when identity is not verified by a strong credential for unique identification (such as holder-of-key in a PKI, or a credential derived from that credential). The PKI usage is so fundamental to this approach that we have provided non-certificated users a way to obtain a temporary PKI certificate based upon their enterprise need and the level of identity assurance needed to provide access and privilege to applications. The process is fully compatible with ELS and works as a complement to existing infrastructure. This work is part of a body of work for high-assurance enterprise computing using web services. Elements of this work are described in [16–22]. This work has raised a number of issues as well as identifying primary capabilities. First among these are the number and types of hardware and protocols that will be supported. Work has begun on the layer 7 services necessary to implement an Ad Hoc networking capability while maintaining the high level of security in ELS. No firm date for implementation has been established, but a target date for capabilities demonstration is in the 2020–2021 time frame.

Acknowledgment. This work was supported in part by the U.S. Secretary of the Air Force and The Institute for Defense Analyses (IDA). The publication of this chapter does not indicate endorsement by any organization in the Department of Defense or IDA, nor should the contents be construed as reflecting the official position of these organizations.

References

1. MANET Definition. http://techterms.com/definition/manet
2. Simpson, W.R.: Enterprise Level Security – Securing Information Systems in an Uncertain World, by Auerbach Publications, May 2016. CRC Press. ISBN 9781498764452. 397 p
3. Simpson, W.R., Foltz, K.E.: Mobile Ad-hoc for Enterprise Level Security. In: Proceedings of The World Congress on Engineering and Computer Science. Lecture Notes in Engineering and Computer Science, 23–25 October 2018, San Francisco, USA, pp. 172–177 (2018)
4. Simpson, W.R., Foltz, K.E.: Enterprise End-Point Device Management. In: Proceedings of The World Congress on Engineering 2018. Lecture Notes in Engineering and Computer Science, 4–6 July 2018, London, U.K., pp. 331-336 (2018)
5. Simpson, W.R., Foltz, K.E.: Enterprise End-point Device Management. In: Process, Proceedings of the World Congress on Engineering. Lecture Notes in Engineering and Computer Science, July 2018, Imperial College, London, pp. 331–336 (2018)
6. Ghemawat, S., Gobioff, H., Leung, S.-T.: The Google file system. In: SOSP (2003)
7. Graefe, G.: Query evaluation techniques for large databases. ACM Comput. Serv., **25**(2) (1993)
8. Hammerbacher, J.: Managing a large Hadoop cluster. Presentation, Facebook Inc., May 2008
9. Mishra, P., Eich, M.H.: Join processing in relational databases. ACM Comput. Serv. **24**(1) (1992)
10. Hardware security module. Wikipedia, the free encyclopedia. http://en.wikipedia.org/wiki/Hardware_security_module
11. Common Access Card (CAC). http://www.cac.mil/common-access-card/
12. Trusted Platform Module (TPM) Summary. http://www.trustedcomputinggroup.org/trusted-platform-module-tpm-summary/
13. Olston, C., Reed, B., Srivastava, U., Kumar, R., Tomkins, A.: Pig latin: a not-so-foreign language for data processing. In: SIGMOD, pp. 1099–1110 (2008)
14. Pavlo, A., Paulson, E., Rasin, A., Abadi, D.J., Dewitt, D.J., Madden, S., Stonebraker, M.: A comparison of approaches to large-scale data analysis. In: SIGMOD (2009)
15. Schneider, D.A., DeWitt, D.J.: A performance evaluation of four parallel join algorithms in ashared-nothing multiprocessor environment. In: SIGMOD (1989)
16. Simpson, W.R., Foltz, K.E.: Enterprise level security: insider threat counter-claims. In: Proceedings of The World Congress on Engineering and Computer Science. Lecture Notes in Engineering and Computer Science, 25–27 October 2017, San Francisco, USA, pp. 112–117 (2017)
17. Simpson, W.R., Foltz, K.E.: Proceedings of the 22nd International Command and Control Research and Technology Symposium (ICCRTS), "Escalation of Access and Privilege with Enterprise Level Security," Los Angeles, CA, pp. TBD, September 2017
18. Simpson, W.R., Foltz, K.E.: Proceedings of the 19th International Conference on Enterprise Information Systems (ICEIS 2017), Porto, Portugal, 25–30 April, 2017, "Enterprise Level Security with Homomorphic Encryption," SCITEPRESS – Science and Technology Publications, vol. 1, pp. 177–184 (2017)
19. Foltz, K., Simpson, W.R.: Enterprise considerations for ports and protocols. In: Proceedings of The World Congress on Engineering and Computer Science. Lecture Notes in Engineering and Computer Science, 19–21 October 2016, San Francisco, USA, pp. 124–129 (2016)
20. Foltz, K.E., Simpson, W.R.: Simplified key management for digital access control of information objects. In: Proceedings of The World Congress on Engineering. Lecture Notes in Engineering and Computer Science, 29 June–1 July 2016, London, U.K., pp. 413–418 (2016)

21. Foltz, K.E., Simpson, W.R.: Proceedings of The 20th World Multi-Conference on Systemics, Cybernetics and Informatics: WMSCI, "Enterprise Level Security – Basic Security Model," Volume I, WMSCI 2016, Orlando, Florida, 8–11 March 2016, pp. 56–61 (2016)
22. Foltz, K.E., Simpson, W.R.: Wessex Institute, Proceedings of the International Conference on Big Data, BIG DATA 2016, "Access and Privilege in Secure Big Data Analysis," 3–5 May 2016, Alicante, Spain, pp. 193–205 (2016)

The Role of Internet of Things and Digital Twin in Healthcare Digitalization Process

Carlotta Patrone[1(✉)], Marco Lattuada[2], Gabriele Galli[3], and Roberto Revetria[3]

[1] General Directorate, E.O. Galliera Hospital, Mura delle Cappuccine 14, Genoa, Italy
carlotta.patrone@galliera.it

[2] Department of Anaesthesiology, E.O. Galliera Hospital, Mura delle Cappuccine 14, Genoa, Italy
marco.lattuada@galliera.it

[3] Department of Mechanical Engineering, Energy, Management and Transports (D.I.M.E.), University of Genoa, Genoa, Italy
gabrigalli95@gmail.com, roberto.revetria@unige.it

Abstract. Among all the work and organization contests, hospitals represent one of the most complex systems and are difficult to manage because unpredictable or highly variable elements coexist in daily routine. In fact, variability plays a key role in the healthcare sector. Operating Rooms (OR) are one of the most crucial environments in the hospital and the way they are run affects many other processes. Medical staff has to note down the various times that represent all the states that the patient has to undergo to successfully conclude the surgery. These procedures are normally made either at the end of the shift or anytime the staff can find some free time. Undeniably, this may lead to human mistakes and inaccuracies. The authors described the application of tools and methodologies borrowed from Industry 4.0 (IoT, Digital Twin, Decision Support System (DSS), Data Mining and System Dynamics) in order to explore better and, possibly improve, the efficiency of ORs. During an 18 days period in June 2018, real time data acquisition was performed in 30 surgeries in an Italian hospital by an independent observer not involved in the patient treatment while the patient's care providers (anesthesiologists, surgeons, nurses) were unaware of the activity of the observer. The authors demonstrated, through ANOVA analysis, that the use of Internet of Things is a powerful instrument for real time data caption and for avoiding artificial variability.

Keywords: Data Mining · Decision Support System (DSS) · Digital Twin · Healthcare · Hospital · Industry 4.0 · Internet of Things (IoT) · Real time · System Dynamics

1 Introduction

Among all the work and organization contests, hospitals represent one of the most complex systems and are difficult to manage because unpredictable or highly variable elements coexist in daily routine [1, 2].

S.-I. Ao et al. (Eds.): WCECS 2018, *Transactions on Engineering Technologies*, pp. 30–37, 2020.
https://doi.org/10.1007/978-981-15-6848-0_3

In fact, variability plays a key role in the healthcare sector [3]: it occurs in the hospital in different forms such as clinical variability, flow variability and professional variability [3]. In addition, the mission of the hospital is to provide healthcare and activities focused on diseases control, pain relief, diagnostic activities etc. and, for this reason, often the patient's data entry are not in real time. As a direct consequence, a gap between the real and registered data occurs, besides, these latter are the ones usually used for performance analysis. The evolving technology provides new instruments to detect data in real time.

Operating Rooms (OR) are one of the most crucial environments in the hospital and the way they are ruled affects many other processes such as bed assignment, surgery waiting lists, staff recruitment and so forth [4]. For this reason, optimizing and improving the OR efficiency is one of the most recurring themes in many scientific fields like engineering, medicine, economics and management. Some of the main issues which underlies mistakes and wastes of time are those related to repetitive and manual tasks [5].

Indeed, these activities are frequent in the OR, in fact, medical staff has to note down the various times that represent all the states that the patient has to undergo to successfully conclude the surgery. These procedures are normally made either at the end of the shift or anytime the staff can find some free time. Undeniably, this may lead to human mistakes and inaccuracies [5].

In the following chapter, the authors will describe the possible application of tools and methodologies borrowed from Industry 4.0 (IoT, Digital Twin, Decision Support System (DSS), data mining, system dynamics) in order to explore better and, possibly, improve the efficiency of ORs.

2 Material and Methods

2.1 Terms and Definition

The IEEE Community defines the IoT as: "… a selfconfiguring and adaptive system consisting of networks of sensors and smart objects whose purpose is to interconnect "all" things, including every day and industrial objects, in such a way as to make them intelligent, programmable and more capable of interacting with humans" [6].

The concept of Digital Twin has been introduced in 2002 by Professor Michael Grieves at University of Michigan. He defines Digital Twin as: "the Digital Twin is a set of virtual information constructs that fully describes a potential or actual physical manufactured product from the micro atomic level to the macro geometrical level. At its optimum, any information that could be obtained from inspecting a physical manufactured product can be obtained from its Digital Twin" [7].

One of the first contextualization and definition of DSS was provided by Sprangue back in 1980. He said that: "The concepts involved in DSS were first articulated in the early '70's by Michael S. Scott Morton under the term "management decision systems"… A few firms and scholars began to develop and research DSS, which became characterized as interactive computer-based systems, which help decision makers utilize data and models to solve unstructured problems" [8].

The authors defined Data Mining as "the science of extracting useful knowledge from such huge data repositories." or "[…] the science of extracting useful knowledge from such huge data repositories." Furthermore, they affirm that Data Mining is "a broad field that combines techniques from different areas in computer science and statistics" [9].

The authors pointed out that: "system dynamics involves the ability to represent and assess the dynamic complexity of the behavior that arises from the interaction of a system's agents over time both textually and graphically. Plus, it is also a set of synergistic analytic skills used to improve the capability of identifying and understanding systems, predicting their behaviors, and devising modifications to them in order to produce desired effects. These skills work together as a system" [10].

2.2 Literature Selection

In the literature, IoT applications in the healthcare sector can be found with a particular focus on the Operating Room [11]. IoT can be used with different aims in OR such as to improve the technical skills of the young orthopaedics [12], for anhestesia data caption [13] or for govern the drugs along with the supply chain and for the patient safety in the OR [14]. Gomes C et al. reported a new Decision Support System supported by Data Mining and Simulation tecniques for the surgery in a Portuguese hospital which aims to help the staff to manage the patients' scheduling and resource allocation processes [15]. All of these fields converge in Industry 4.0. A literature review on Healthcare 4.0 has been carried out [16]. C. Patrone et al showed an interesting application of IoT in a smart post-hospitalization for older people [17].

2.3 Case Study

As previously stated, a possible difference between the real time of an event and the corresponding registered one may differ, even significantly, affecting any inference on the data themselves. This bias can be highlighted, for example, coupling data registration on medical records during the surgical process in the OR with a real time acquisition using IoT.

The case study introduced consists of a real time data driven Digital Twin (DT) modeled in System Dynamics and enabled by IoT buttons. This type of buttons is small, easy to use and, once pressed, it sends a per-programmed message through a network to a server. In the setting here described, two different AWS (Amazon Web Service) IoT buttons are used and once pressed, they record treatment time and then transmit it to a database used for data visualization and analysis. Furthermore, they become the input data of a System Dynamics model which outlines the operation flow.

During an 18 days period in June 2018, real time data acquisition was performed in 30 surgeries in an Italian hospital by an independent observer not involved in the patient treatment while the patient's care providers (anesthesiologists, surgeons, nurses) were unaware of the activity of the observer. The following two types of routine elective surgeries have been considered:

- *ordinary surgery:* pre-booked and scheduled in the ordinary surgery activity and admitted to the ward;

– *day surgery:* pre booked and scheduled in the ordinary surgery activity without admission to the ward, here after the anesthesia recovery, the patient is ready to go home within two days.

Before surgery, the patients are submitted to a pre-surgery visit. Here, they are classified as ordinary or day surgery, according to the type the treatment they need. This is a pivotal decision since it is related to the type of operation and to the availability of hospital beds.

Nine instants describe the patient pathway, corresponding to a punctual time (HH:MM):

- T0-patient gets into the Block of Operating Rooms (BOR)
- T1-patient gets into the OR
- T2-anesthesia infused, patient ready for surgery
- T3-skin incision
- T4-skin suture
- T5-end of anesthesia
- T6-exit from the OR
- T7-exit from the BOR
- T8-OR is available for next surgery

For each instant an external observer, a student, pushed the button. The device is connected to Wi-Fi and it directly collects the data in a database implemented in Excel.

Several methods are available for comparing the two data sets in order to demonstrate the effectiveness of this new approach. The easiest one is to calculate the difference between the time Tn collected with IoT buttons and the same time Tn collected without the IoT button. For statistical analysis, two-way Analysis of Variance (ANOVA) was performed. Table 1 demonstrates the results of two-way ANOVA.

2.4 Digital Twin

A Digital Twin has been created with Powersim produced by Powersim Software AS, this model is continuous and it has been created using a System Dynamics approach. It reproduces the patient's event-chain starting from their arrival in the OR to their exit. This model exploits a top-down approach where each state represents one of the nine states (T1-T9) and it depends on the previous state. Therefore, once the time of the previous event has been completed, the flow is enabled and the patient can move to the next state. The change of a state is triggered by a condition that allows the patient's flow across the arrows from one state to another.

Data collected from the buttons are stored in a database that are used to feed the Digital Twin which simulates surgery processes. Thanks to this system, knowing the expected waiting time and the probability to be operated in an established time becomes possible. This is useful both for patients who will be able to know how long they will have to wait for their surgeries, and even for nurses and surgeons who will be able to schedule a better and leaner operation planning.

Table 1. Two – Way ANOVA results

	Times	Patient 1	Patient 2	Patient 3	Patient 4	Patient 5	Patient 6	Y_i
T_1–T0	**IoT**	0.90	17.22	6.13	28.10	0.10	16.00	68.45
	Manual	8.00	0.00	0.00	50.00	7.00	6.00	71.00
			SS	**DoF**	**MS**	$\mathbf{F_0}$	$\mathbf{F_{tab}}$	Y = 139.45
		SSTratt	0.54	1	0.54	0.0025	4.96	
		SSE	2152.38	10	215.24			
		SST	2152.92	11		**SIGNIFICANT**		
T_2–T_1	**IoT**	2.95	36.32	0.13	55.88	1.90	53.00	150.18
	Manual	26.00	10.00	10.00	11.00	17.00	49.00	123.00
			SS	**DoF**	**MS**	$\mathbf{F_0}$	$\mathbf{F_{tab}}$	Y = 273.18
		SSTratt	61.58	1	61.58	1.74E−27	4.96	
		SSE	3.53E+29	10	3.53E+28			
		SST	3.53E+29	11		**SIGNIFICANT**		
T_3–T_2	**IoT**	51.83	28.70	45.88	1.08	3.73	8.00	139.23
	Manual	23.00	33.00	10.00	4.00	17.00	10.00	97.00
			SS	**DoF**	**MS**	$\mathbf{F_0}$	$\mathbf{F_{tab}}$	Y = 236.23
		SSTratt	148.64	1	148.64	1.04E−06	4.96	
		SSE	1.424E+09	10	1.42E+08			
		SST	1.424E+09	11		**SIGNIFICANT**		
T_4–T_3	**IoT**	23.10	1.40	1.13	25.62	10.02	18.00	79.27
	Manual	19.00	32.0	34.00	25.00	30.00	15.00	155.00
			SS	**DoF**	**MS**	$\mathbf{F_0}$	$\mathbf{F_{tab}}$	Y = 234.27
		SSTratt	477.96	1	477.96	4.50E−33	4.96	
		SSE	1.062E+36	10	1.06E+35			
		SST	1.06E+36	11		**SIGNIFICANT**		
T_5–T_4	**IoT**	1.70	3.72	3.08	10.27	60.23	2.00	60.75
	Manual	0.00	5.00	6.00	0.00	1.00	1.00	53.00
			SS	**DoF**	**MS**	$\mathbf{F_0}$	$\mathbf{F_{tab}}$	Y = 113.35

(*continued*)

Table 1. (*continued*)

	Times	Patient 1	Patient 2	Patient 3	Patient 4	Patient 5	Patient 6	Y_i
		SSTratt	12.58	1	12.58	9.55E−06	4.96	
		SSE	1.32E+07	10	1.32E+06			
		SST	1.32E+07	11		**SIGNIFICANT**		
T_6–T_5	**IoT**	15.13	6.20	34.70	1.95	2.72	6.00	66.70
	Manual	8.00	0.00	5.00	16.00	9.00	7.00	45.00
			SS	**DoF**	**MS**	$\mathbf{F_0}$	$\mathbf{F_{tab}}$	Y = 111.70
		SSTratt	39.24	1	39.24	6.93E−07	4.96	
		SSE	1.62E+05	10	1.62E+04			
		SST	1.63E+05	11		**SIGNIFICANT**		
T_7–T_6	**IoT**	12.47	12.53	9.33	14.30	3.12	9.00	60.75
	Manual	11.00	20.00	5.00	0.00	8.00	9.00	53.00
			SS	**DoF**	**MS**	$\mathbf{F_0}$	$\mathbf{F_{tab}}$	Y = 145.30
		SSTratt	4.50	1	4.50	6.93E−07	4.96	
		SSE	6.50E+07	10	6.50E+06			
		SST	6.50E+07	11		**SIGNIFICANT**		
T_8–T_7	**IoT**	13.13	19.05	12.80	8.02	15.30	8.00	76.30
	Manual	9.00	10.00	15.00	14.00	10.00	11.00	69.00
			SS	**DoF**	**MS**	$\mathbf{F_0}$	$\mathbf{F_{tab}}$	
		SSTratt	4.44	1	4.44	4.93E−18	4.96	
		SSE	9.02E+18	10	9.01E+17			
		SST	9.02E+18	11		**SIGNIFICANT**		

3 Discussion

The healthcare sector is affected by natural and artificial variability [3]. The natural variability is random and it is the result of the intrinsic element of health care delivery. An example of natural variability is that every patient is different than the other because of age, co-morbidities, response to therapy and so on. The artificial variability is not random and it is usually linked with defects or with wrong decisions made by the organization. An example of artificial variability made by Litvak [3]: is "the unfamiliarity with a new technology can be eliminated through education and certification". In other words, manager can focus on the topic and eliminate the artificial variability whereas the natural variability can only be observed and measured.

These general principles are well shown in the table where ANOVA analysis reports through SSE a high level of noise. A limitation of the study is due to the sample size and to the surgical operations selected: a single surgical specialty has been selected (orthopedics) and not a single type of surgery (i.e. hip replacement).

The need to explore tools that allow clinicians to better schedule the surgical activity avoiding the artificial variability influence is still under explored. For example, Gomes et al. used the DSS to optimize surgery schedules and to measure their quality in order to reduce the waiting lists [15].

More recently Karakra et al. used the definition of Digital Twin in the healthcare sector. In detail, they used IoT Devices for real time data caption [18].

This data feed the Discrete Event Simulation model implemented through FlexSim, produced by FlexSim Software Products, Inc. The system implemented by the authors simulates the efficiency of the services according to which the decision maker can adapt the activities schedule (labs, diagnostic procedures, etc.). More details about the use of Digital Twin in healthcare can be found in [19] and [20].

Starting from this background the authors decided to evolve the use of real time simulation developing a tool daily available for the decision maker in surgeries. This implies the possibility to obtain a suggestion from the model to a better allocation of the resources. For example, the model could be able to predict, based on a datawarehouse of similar surgeries, the expected time of the end of the surgical act (T4) so that the decision maker is able to optimize how to procede according to the total time available for surgery in a single day. Better use of available resource is the direct consequence.

4 Conclusion

The authors strongly believe, according to Litvak, that to measure processes avoiding artificial variability is a key point for improvement. The use of Internet of Things is a powerful instrument to perceive this goal as shown in this work.

Acknowledgment. The authors would like to thank Alireza Khodabakhsh for his contribution.

References

1. Patrone, C., Cassettari, L., Damiani, L., Mosca, R., Revetria, R.: Optimization of lean surgical route through POCT acquisition. In: Proceedings of the International MultiConference of Engineers and Computer Scientists 2017 vol II, IMECS 2017 (2017)
2. Patrone, C., Lagostena, A., Revetria, R.: Managing and evaluating different projects in a hospital trough the analytic hierarchy process: methodology and test case. In: IEEE International Conference on Industrial Engineering and Engineering Management (IEEM) 2017, pp. 894–898 (2017)
3. Litvak, E., Long, M.: Cost and quality under managed care: irreconcilable differences. Am. J. Manag. Care **6**(3), 305–312 (2000)
4. Chang, J., De Carvalho, T., De Souza Santos, S., Da Silva Fernandes, A.: Operating rooms optimization in a cardiology public school hospital: the joint and sequential use of the models of Min and Beliën. In: PICMET 2016 - Portland International Conference on Management of Engineering and Technology: Technology Management for Social Innovation, Proceedings, 3106 p. (2017)

5. Demartini, M., Damiani, L., Patrone, C., Revetria, R., Tonelli, F., Giribone, P.: Internet of Things in healthcare system: from theoretical investigation to practical implementation, paper submitted
6. Laplante, P.A., Kassab, M., Laplante, N.L., Voas, J.M.: Building caring healthcare systems in the Internet of Things. IEEE Syst. J. **12**(3), 3030–3037 (2018)
7. Grieves, M., Vickers, J.: Digital twin: mitigating unpredictable, undesirable emergent behavior in complex systems. In: Kahlen, F.-J., Flumerfelt, S., Alves, A. (eds.) Transdisciplinary Perspectives on Complex Systems, pp. 85–113. Springer, Cham (2017). https://doi.org/10.1007/978-3-319-38756-7_4
8. Sprague, R.: A framework for the development of decision support systems. MIS Quart. Manage. Inf. Syst. **4**, 1–26 (1980)
9. Chakrabarti, S., Ester, M., Fayyad, U., Gehrke, J., Han, J., Morishita, S., Piatetsky-Shapiro, G., Wang, W.: Data mining curriculum: a proposal (Version 1.0). In: Intensive Working Group of ACM SIGKDD Curriculum Committee (2006)
10. Rahim, F., Hawari, N., Abidin, N.: Supply and demand of rice in Malaysia: a system dynamics approach. Int. J. Supply Chain Manag. **6**(4), 234–240 (2017)
11. Patrone, C., Khodabakhsh, A., Lattuada, M., Revetria, R.: Internet of Things application in the healthcare sector. In: Proceedings of the World Congress on Engineering and Computer Science 2018, 23–25 October, 2018, San Francisco, USA, pp. 449–452. Lecture Notes in Engineering and Computer Science (2018)
12. De La Borbolla, I.R., Chicoskie, M., Tinnell, T.: Applying the Internet of Things (IoT) to biomedical development for surgical research and healthcare professional training. In: IEEE Technology and Engineering Management Society Conference, July 2017, pp. 335–341 (2017)
13. Kadry, B., Feaster, W., MacArio, A., Ehrenfeld, J.M.: Anesthesia information management systems: past, present, and future of anesthesia records. Mt. Sinai J. Med. **79**(1), 154–165 (2012)
14. Paulin, A., Thuemmler, C.: Dynamic fine-grained access control in e-health using: the secure SQL server system as an enabler of the future internet. In: 2016 IEEE 18th International Conference on e-Health Networking, Applications and Services, Healthcom 2016 (2016)
15. Gomes, C., Sperandio, F., Borges, J., Almada-Lobo, B., Brito, A.: A decision support system for surgery theatre scheduling problems. In: Cruz-Cunha, M.M., Varajão, J., Powell, P., Martinho, R. (eds.) CENTERIS 2011. CCIS, vol. 221, pp. 213–222. Springer, Heidelberg (2011). https://doi.org/10.1007/978-3-642-24352-3_23
16. Patrone, C., Cassettari, L., Saccaro, S.: Industry 4.0 and its applications in the healthcare sector: a sistematic review. In: Submitted to the XXIV Edition of the Summer School Francesco Turco-Industrial Systems Engineering 2019. AIDI-Italian Association of Industrial Operations Professors (2019)
17. Patrone, C., Cella, A., Martini, C., Pericu, S., Femia, R., Barla, A., Porfirione, C., Puntoni, M., Veronese, N., Odone, F., Casiddu, N., Rollandi, G.A., Verri, A., Pilotto, A.: Development of a smart post-hospitalization facility for older people by using domotics, robotics, and automated tele-monitoring. Geriatric Care: accepted 2nd of April 2019 (in press)
18. Karakra, A., Fontanili, F., Lamine, E., Lamothe, J., Taweel, A.: Pervasive computing integrated discrete event simulation for a hospital digital twin. In: Proceedings of IEEE/ACS International Conference on Computer Systems and Applications, AICCSA 2019 (2019)
19. Patrone, C., Galli, G., Revetria, R.: A state of the art of digital twin and simulation supported by data mining in the healthcare sector. In: Frontiers in Artificial Intelligence and Applications, vol. 318, pp. 605–615 (2019)
20. Galli, G., Patrone, C., Bellam, A.C., Annapreddy, N.R., Revetria, R.: Improving process using digital twin: a methodology for the automatic creation of models. In: Lecture Notes in Engineering and Computer Science, pp. 396–400 (2019)

Big Data Based VOC for Vehicle Development

Yuan Li(✉) and Jiaying Lv(✉)

Beijing Motor Company, No. 99 Shuanghe Street, Renhe Town, Beijing, Shunyi District, China
18500293291@163.com, lvjiaying@beijing-atc.com.cn

Abstract. Although there are many design processes and design methodology progresses in automotive development in the last few decades by using computer technology such as CAD, CAE, PDM, PLM and etc., one the real gaps between customers and the vehicle designers still exists. That is, the vehicle OEMs still don't know what exactly customers want and feel about their products (we usually call these VOC) in real time and how to meet those demands. We used tools like VOC, QFD, CRM, marketing research, dealer visiting, warranty database, but all these tools and data lack of real time, data infidelity, small sampling pool and high cost. In this paper, we present a technology that help on closing the gap by using the Internet+ technology or the big data analysis technology. Specially, we utilized and advanced the big data analysis technology to help the engineers on vehicle system design rather than a high level public sentiment assessment. The process and the computer system we developed are also described.

Keywords: Automotive · Data acquisition · Data analysis · VOC · Text mining · Information extraction

1 Introduction

Although there are many design processes and design methodology progresses in vehicle development in the last few decades by using computer technology such as CAX, PDM, BOM, PLM, CRM and etc., one the real gaps between customers and the vehicle designers still exists. That is, the vehicle OEMs still don't know what exactly customers want and need about their products in real time and how to meet their demands. We used tools like VOC, QFD, CRM, marketing research, dealer visiting, warranty database, but all these tools and data are lack of real time, data infidelity, small sampling pool and high cost. They require manual intervention and hardly an automated process.

Nowadays, however, with the advancement of the internet and mobile technology, it's the habit of vehicle buyers and consumers to express their comments including unsatisfaction and needs on various vehicles, show off their vehicles and most of the time publish their complains and needs about their vehicles in the internet sites such as Autohome (40% market share of the vehicle online shopping sites in China), Yiche, Renrenche, Weibo (Microblog), Wechat, Youtube and etc. in the format of text and photos in chat rooms, user group discussion rooms and etc. With close integration to

Y. Li—Member of IAENG, SAE, and CAA.

S.-I. Ao et al. (Eds.): WCECS 2018, *Transactions on Engineering Technologies*, pp. 38–47, 2020.
https://doi.org/10.1007/978-981-15-6848-0_4

these websites, we can further dig down for more specific information such as buying intents and budget. Today, we simply call them cloud information. This information is valuable to OEM if obtained. It contains indefinite information about the car quality feedback and complains of existing buyers, the appearance and styling that delighted the shoppers, option preference and etc. The information can be very specific to certain model and reflects the true feeling. When used well, it can not only alert the OEMs about the immediate quality crisis and malfunctions but also be used to identify future vehicle trends. It's valuable information for vehicle planning people and engineers for their future design and the vehicle annual upgrades. It can be used for most of vehicle development phases such as strategic intention, strategic concept and all the way to design frozen as show in Fig. 1 [1]. The cloud information has the characteristics of large sample pool, fast, cost free and genuine. Traditionally, we need spend time and money to gather these information from vehicle yearbooks, industry reports, company internal analysis reports, warranty database and etc. however, most of the time, these information are small in sample quantity and slow in response time. Therefore, how to use well the cloud data may actually define how close companies to their customers and the future of these companies. In fact, based on our experience, it also can point the direction that is hard to provide in traditional process and sometimes presented as a surprise to the company.

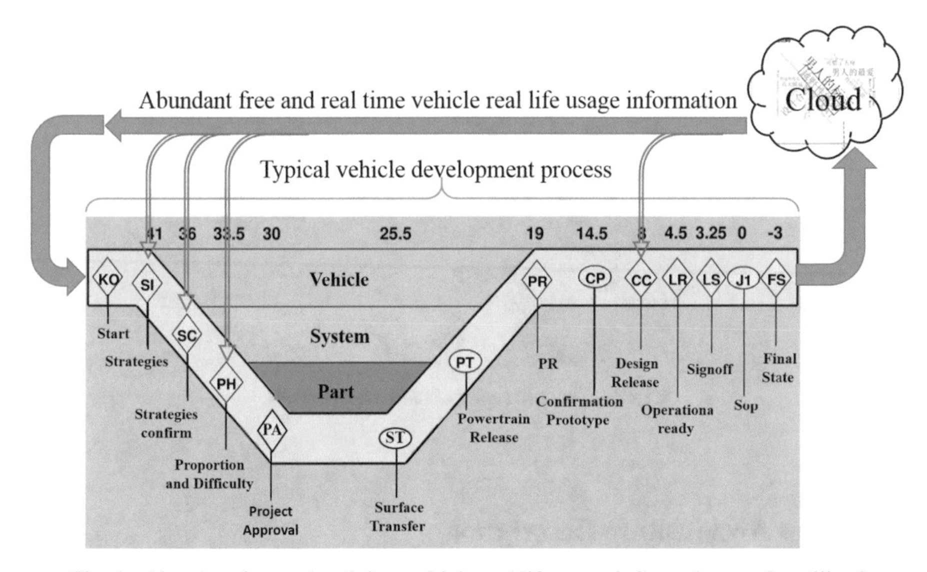

Fig. 1. Abundant free and real time vehicle real life usage information can be utilized

In this paper, we present our efforts and methods on utilizing the data. Specially, we utilized and advanced the big data analysis technology to help the engineers on vehicle system design rather than only a high level public sentiment assessment. The process is almost an automatic process and can be integrated with other computer systems such as ECM (engineering change management, BOM (Bill of Material), PDM (Product Data Management) and etc.

By the integration of these systems, it not only streamlined the design change process but also provided the opportunities to utilize the Linked Data property in data in these systems which strengthened the Key Words searching scope. The existing systems in general exchange data in an encrypted data stream, this requires the data be decrypted process when systems were integrated.

In our effort on this topic started 4 years ago, we found then that how to effectively do the text mining and data cleansing in auto industry is still in its infancy and there are no mature software system, process and successful business examples or in other words, the accuracy of the results are not able to reach to a satisfaction level. Based on this finding, we decided to develop our own system and algorithm. This system focused on engineer's daily work and enabled the engineers to design the vehicle and its subsystems to meet custom needs. We called this function as "deep guidance" and we also believed this is a paradigm change both in a company's development process as well as the information system architecture. Rather than getting the striped down information through marketing and planning department of the company in the traditional or over-the-wall way, the engineers can access these data first hand which help them understand the company's vehicle planning and option decision at least as shown in Fig. 2.

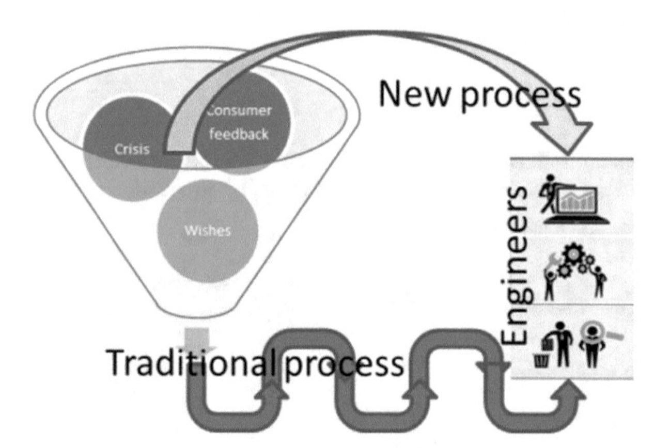

Fig. 2. New process vs. traditional process

2 System Architecture Description

The system developed has three main layers: data collection, data analysis, and result presenter as shown in Fig. 3. The data collection layer has six main modules. They are, data planning, data acquisition, information extraction, data warehousing, text mining, and content analysis and management. According our company vehicle style and future direction, we plan and define the data we want to crawl in the data planning module and then, the distributed multi-thread data acquisition module automatically crawls the data to our database. In this way, we optimized the data collected to be the most valuable asset rather than all the data from internet. We have configured the system to crawl data from 200+ automotive related sites where posts containing text and photos about vehicle information and feedback running multi-million lines daily. Data from company owned sites and contracted sites were also merged with crawled data in the data warehouse module and these data has more data dimensions then the crawled data. In the data acquisition, we crawled data such as the industrial trends, benchmark vehicle data, government regulation updates, customization trends, and public sentiments. It supports data formats of Words, PDF, PPT, Excel, CAJ and etc. And also, in the data acquisition module, the method to penetrate the "data block mechanism" has been implemented for each individual site so we can maximize the data volume and it supports RSS sites and agents. The geological data of the publisher has been identified by the IP address so we have a geological distribution of the data feedback. In the database management, we used cluster technology to distribute the data load thus insure the data process speed.

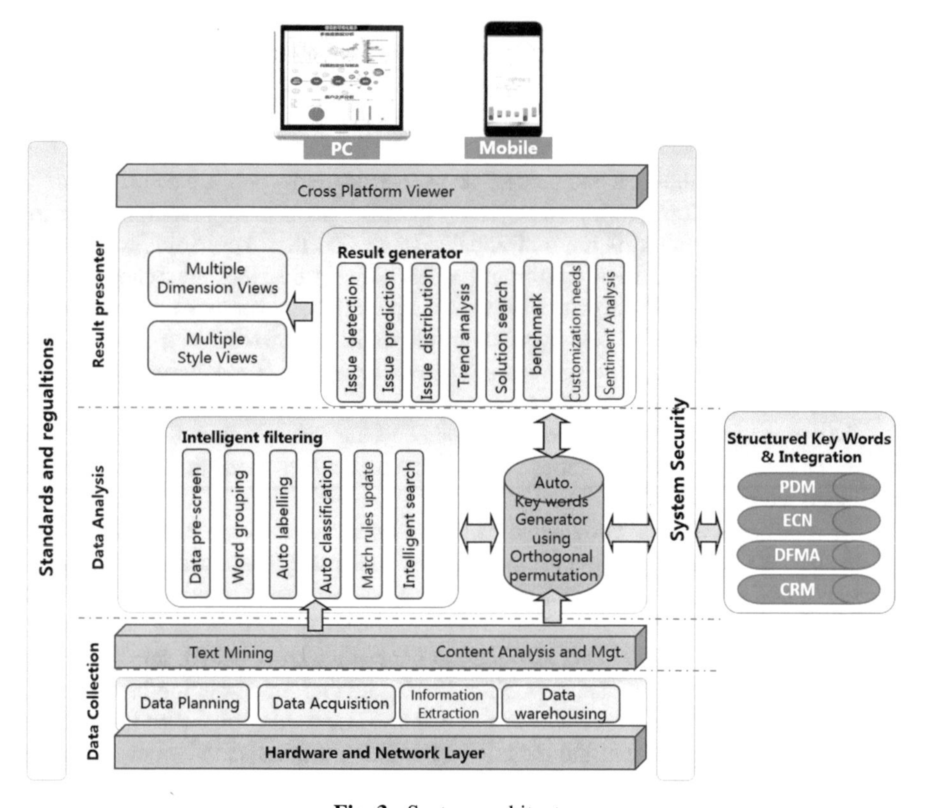

Fig. 3. System architecture

The cross platform viewer use the login information to identify user and personalize his display and match his specialty and interests so that the system only notify the related events and warnings and it supports both PC and smart phone usage.

Once the data is our local database, the information extraction module then use our NLP (nature language processor) analyzer and concept relation dictionary to extract necessary information from the text about the vehicle. The extracted data are stored alongside with internal vehicle quality data and warranty data in the data warehouse. Then, the text mining module matches the text find the quality, benchmark and etc. related information in the data management module. The data warehouse was not only served as a company data asset reservoir but also as the bases for fake, spam and duplicated data identification together with intelligent filtering module where modified factor graph model were implemented [2].

The data analysis layer has two main modules: Intelligent filtering and key word generation. The intelligent filter has the functions of data and user ID pre-screen, word grouping and segmentation, automatic labeling and automatic classification, matching rule upgrades, and intelligent search. The fake and fabricated data and users were detected and excluded in these modules. The searching speed can reach 2100 GB/s. The similarity detection can reach one million article/20 ms. The maximum data table can contain 4 billion entries and the maximum data capacity is 8 TB. The word grouping and segmentation module can match words 5 MB/s and full split speed 2 MB/s and the average accuracy is 97.3%. The key word generator utilized the key words defined in the traditional systems such as PDM, ECN, BOM, FEMA, and CRM. By doing the orthogonal permutation, the key words were generated. They form the foundation of the key words due to their well-structured classification and fields. The well-structured words are natural fit to the concept relation dictionary which can greatly increase the matching accuracy. One the advantages of this method is the traditional systems and their key words were maintained and updated regularly by a rigorous process which reflect the new technology and industrial trends. These key words were then wrap with their synonyms and antonyms. The preliminary ontology of these words were also built and we have seen an increase of the searching accuracy in terms of the relevance to the subject. The interface that allows the user to add new key words were also built.

The automatic classification module implements algorithm such as KNN, VSM and Bayes. These key words then cross matches the text mined from the internet in the intelligent search module.

The result presenter has two main modules: Viewer and result presenter. The Viewer can display result in multiple dimension and styles. The result generator has the functions of issue detection, issue prediction, issue distribution, trends analysis, solution search, benchmark, customization needs, and sentiment analysis [3]. In the solution search module, the system searches the company internal knowledge base, paid article and patent sites as well as public sites such as CNKI, Baidu and etc. for a solution based on the geological information and recent events in that area. Once identified then the solution recommendation module sent the most useful solution articles to the engineer. In the benchmark module, the competitor vehicle performance data such as the fuel consumption, EV range and all the way to their sub-systems performance and properties were benchmarked in multiple dimensions such as the cost, reliability and etc. for engineers to view.

System usage: The system can be used by high rank decision makers and engineers in the field of marketing, quality and planning.

1. Used by the decision makers

The data is organized by adopting the Kano model where the consumer's needs are classified in three aspects, basic needs, performance needs, and delighters as shown in Fig. 4. In general, the standard options provided with by most successful vehicles in the market of the same class produced by Chinese OEMs were considered as the basic needs, the corresponding options in production by the joint venture companies are considered as the performance needs, and the concepts published and highly recognized by public in events and shows were considered delighters. In this way, the feedback are aligned with the internal marketing research and planning process. The data and analysis results used by the planning people and decision makers to decide what's options and features should be equipped on the next vehicle. The engineers used these classification to decide the urgency to change for certain quality issues [2].

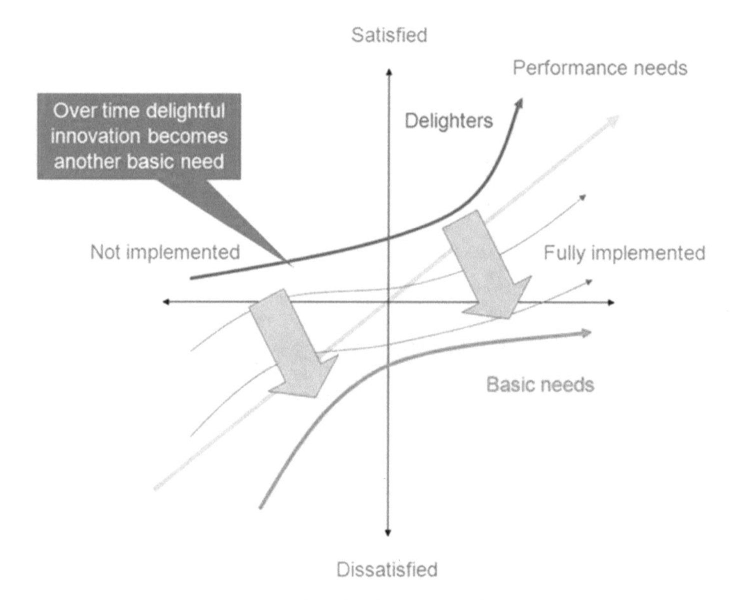

Fig. 4. The Kano model

Take the analysis result of one car which has been already sent to market as an example, the project can search and get VOC of the car any time using the system and decide which could be optimized in the next generation product or in the MY product. Two important charts shows in Fig. 5, Fig. 6 and Fig. 7.

2. Used by the engineers

The engineers who registered in this systems can instantly know the vehicle consumer feedback. The issues are organized in a fish bone diagram and the nodes indicated the root, the vehicle level, system level and sub-system level and component level as shown in the Fig. 8. The competitors' quality level on the sub-systems and components were also displayed to engineers so they can decide what level of improvements are needed with cost consideration.

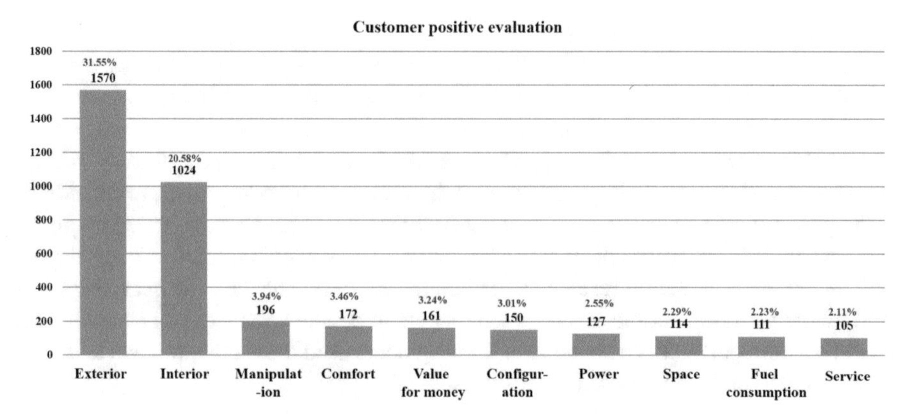

Fig. 5. Customer positive evaluation

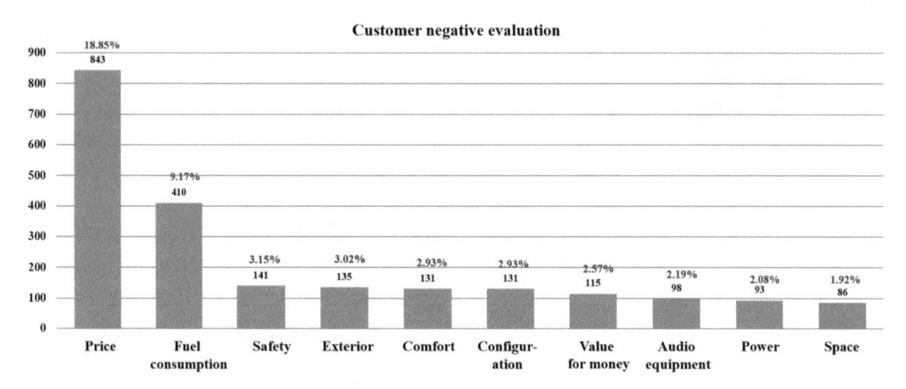

Fig. 6. Customer negative evaluation

The nodes and levels are predefined according to company's Vehicle Partition and Part Structure. It consists of 8 levels and 4897 nodes and the mechanism and process to update both levels and nodes are developed. By clicking on nodes, the engineers then can use the system to dig into the problems. The size of the nodes relatively reflects the number of issues. By click on the node shown "Seat" which is under "IE" which stands for interior and exterior of the vehicle in Fig. 8, for instance, we can find the leather is one of the main issues needs fixed. Once clicked on the leather node, the consumer concerns and photo can be read as shown in Fig. 9.

So far the engineer users spread cross engineering fields including design and release, quality, marketing, customization and aftermarket. The computer screen capture of the node network and bar chart where the number of issues occurred and their distribution in last month are shown in Fig. 10.

The system also provided the users with subscription functionality where specific fields, subsystems and roles can be defined. Once subscribed, the system will notify the engineer of related field on the issue of the car as soon as the consumer complaints about the vehicles in the internet. This minimize the engineer effort to traverse through

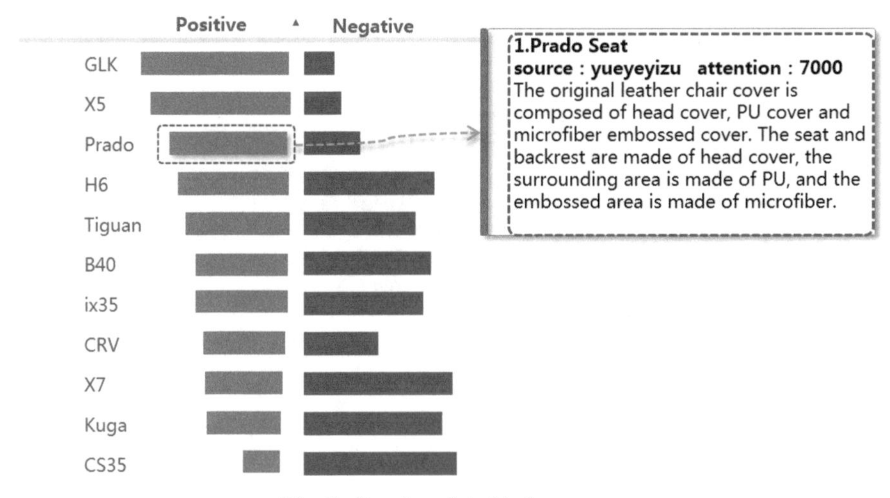

Fig. 7. Benchmark in bigdatasystem

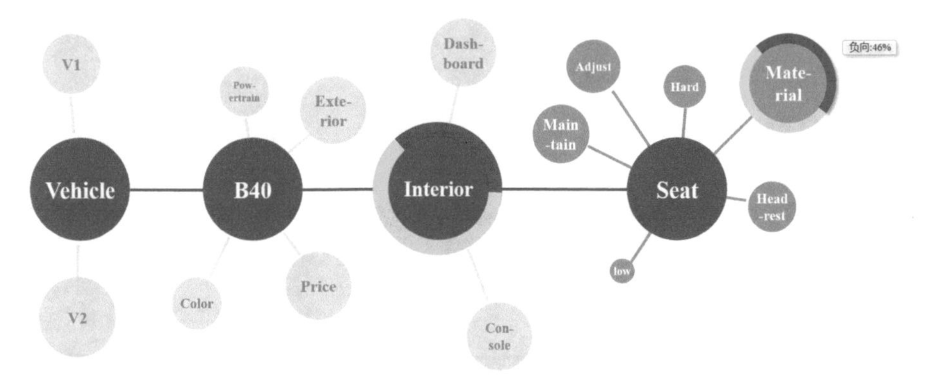

Fig. 8. The issue fish bone network

the issue network [4]. For a pre-defined catalog of urgent problems, the system will automatically initiate an engineering change in the ECM system integrated thus speed up the problem solving process.

3 Next Step

The system still has many improvements pending such as increase the accuracy of the nature language analysis and increase the key word database. The Linked Data property in both internal systems and cloud information will be examined further. Also, we would like to improve the solution recommendation module by sending the most useful solution segments instead of the whole article to the engineer.

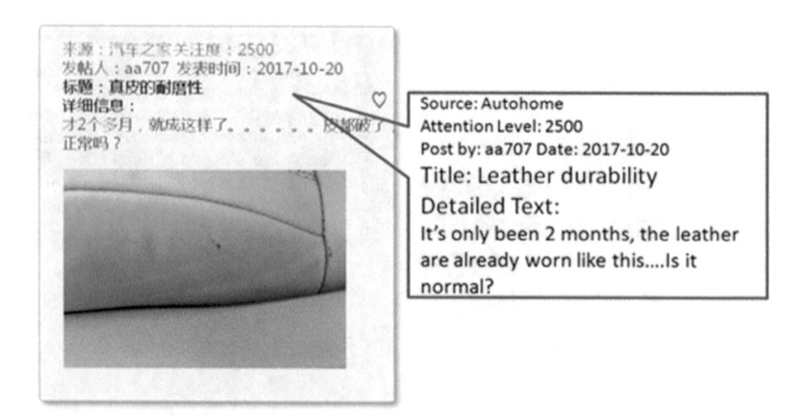

Fig. 9. The issue description published by the consumer where shown the leather worn

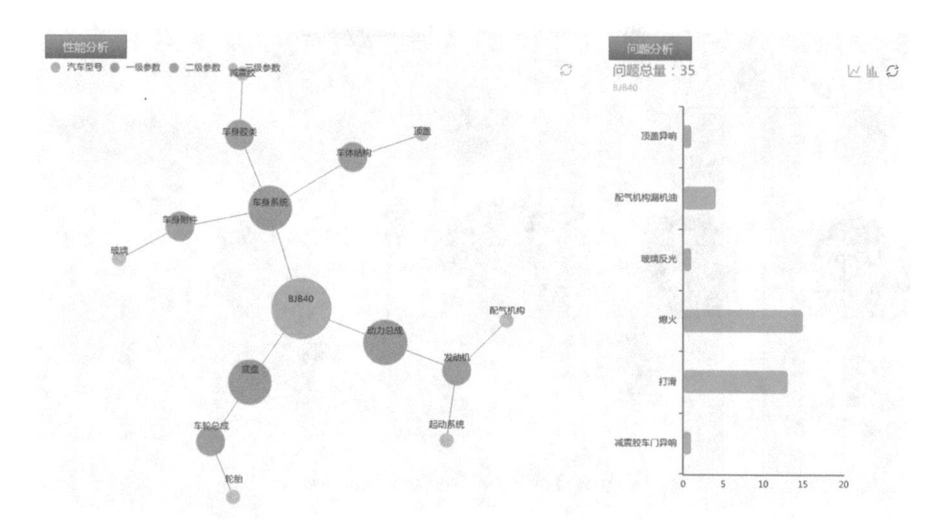

Fig. 10. The number of issues occurred in last month

4 Conclusion

From Oct. 2014 to Dec. 2015, we have successfully developed the big data based intelligent analysis platform for vehicle development. Since then, it went through several upgrades. So far the system has identified 696 issues and guided the vehicle design and planning of 12 vehicle model-years. The user base has reached 3000 inside Beijing Motor Company. It created direct links between vehicle consumers and vehicle designers. It has not only reduced the issue solving cycle time but also prevented critical issues that lead vehicle safety issues and potentially accidents. It also changed the company's IT architecture and became one of the fundamental system in the company.

References

1. Li, Y., Wang, Z., Wang, L., Lv, J.: Close the last gap of R&D in automotive design. In: Proceedings of the World Congress on Engineering and Computer Science, pp 394–397. Lecture Notes in Engineering and Computer Science, 23–25 October, 2018, San Francisco, USA (2018)
2. Guo, Q.J., Jia, M.L., Li, Y.: Research on components' relationship and configuration management for multi model structure. J. New Ind. **5**(5), 62–66 (2015). https://doi.org/10.3969/j.issn.2095-6649.2015.05.09
3. Li, Y.: Increase the PDM data security level by data encryption. J. CAD/CAM Ind. Manuf. **11**, 37–39 (2019)
4. Lu, Y.Q., Zhang, L., Xiao, Y.D., Li, Y.G.: Simultaneously detecting fake reviews and review spammers using factor graph model. In: Davis, H.C., Halpin, H., Pentland, A. (eds.) Proceedings of the 5th Annual ACM Web Science Conference (WebSci 2013). ACM Press, New York (2013)

Safe Management of Complex River Systems

Roberto Revetria[1](✉), Anatoly Aleksandrov[2], Mikhail Ivanov[2], Konstantin Neysipin[2], and Olga Ivanova[2]

[1] University of Genoa, Via All'Opera Pia 15, Genoa, Italy
roberto.revetria@unige.it

[2] Bauman Moscow State Technical University, 2-ya Baumanskaya, 5-1, Moscow, Russia
{rector,mivanov,nyusipin,olgaivanova}@bmstu.ru

Abstract. Nowadays hydroelectricity is one of the leading power generation technologies, which is widely spread all over the World. Despite low ecological impact during normal operation of the hydro power plants in case of emergencies floods can cause severe damages to the environment, built constructions and inhabitants. And this problem is especially important for large rivers with several dams on them. Proper management of complex river systems is especially important. In this Chapter are analyzed current complex river systems management approaches. It is shown, that none of them consider simultaneously rationalization of energy production, maintaining required water level in the river system and minimizing possible damage from seasonal floods at a complex river system. Hence, basing on the literature review a novel approach to describe the system for finding an operational model guide is presented in the Chapter. The developed approach is applied to the mountain region of Valle d'Aosta in North Western Italy. The dynamic modelling is performed using the Powersim simulation tool. Based on the results, a safe balancing may be performed that prevents from uncertainty in storage and water flow with effective utilization and minimum flood occurrence.

Keywords: Water management · System dynamic modelling · Flood control · Reservoir operations · Water resource modeling · Complex river system

1 Introduction

Nowadays a lot of attention is devoted to catastrophic floods. Due to their rarity, they cause significant damage to the national economy. In addition, due to the growing population of the Earth, there is the need to develop new territories that have not historically been inhabited such as potentially floodable areas [1].

According to [2] in last century has been observed a significant increase of number of emergencies at HPPs that has resulted in serious financial, ecological losses and lethalities (Fig. 1).

At the same time, regular floods are much better studied and forecasted, so it seems possible to concentrate on management water release in order to improve the safety of the adjacent territories.

Another problem is the mean sea level increase. In [3] it is shown that it has an exponential trend, which result in the future scenarios' range from 0.66 ft to 6.6 ft in

S.-I. Ao et al. (Eds.): WCECS 2018, *Transactions on Engineering Technologies*, pp. 48–65, 2020.
https://doi.org/10.1007/978-981-15-6848-0_5

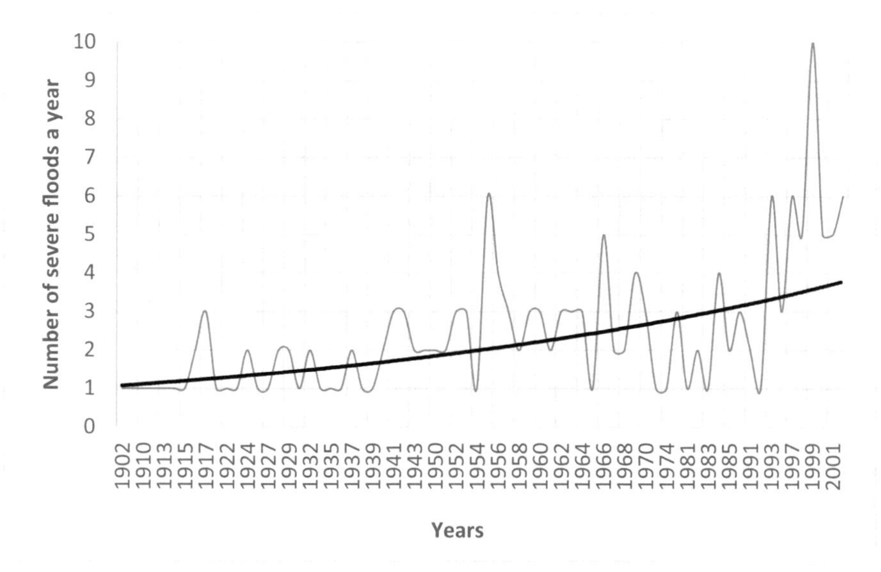

Fig. 1. A trend of severe floods number increase

2100. This, in particular, results in increase of average floods all over the world. Potential for use of social vulnerability assessments to aid decision making for the Colorado dam safety branch [4]. Furthermore, increased population enhances the severity of the flood aftermaths. It is estimated that within 2050 the population will be increased by 130 million, which will demand for a huge increase of water reservoirs.

Complex analysis of the above-mentioned factors results in urgent necessity to develop a safe and reliable algorithms of rivers management. In [5] is provided the list of the most common reasons that resulted in flood emergencies. These are:

- Poor technical condition of the equipment.
- Lack of dam monitoring systems.
- Human factor.
- Change of meteorological conditions.

First three issues may be solved due to proper operation management of the dams. While the last one require a novel systematic approach, that would incorporate state-of-the-art computing, monitoring, forecasting facilities and multi-criteria approach.

2 Mathematical Models for Complex River System Management

Since 1960 various approaches have been developed to use extensively in hydroelectric generation and water management.

Up to now, many approaches to modelling to support water policy management of complex reservoir systems have been proposed. Most of them commonly use a Monte Carlo optimization [6, 7]. It performs optimization over long period of time (several centuries) of historical or synthetically generated discharges. However, optimal management policies may not be applied to any other time series among the one that was calculated. Another widely used approach is based on linear programming, which is often combined with Monte Carlo simulation [8]. The main disadvantage of this method is that all mathematical apparatus has to be linear or linearizable. Another approach is based on the representation of a river network through a set of nodes and links, which is called a network flow optimization. Nodes represent reservoir and links – channels and flows. Such approach is even faster than the others. One of the first attempt to introduce this approach at Missouri river is reported in [9], where it shown that even a simple river basin is very complex to be modelled using this approach when many constraints are considered.

Methods based on linearization may encounter a number of difficulties to be applied precisely due to its large scale or the lack of precision. In these cases a nonlinear programming models might be applied. Nowadays they are considered to be the most advanced due to its power and robustness. However due to its nonlinearity there is always a risk for a model not to converge. Another widely used and well studied approach is dynamic programming [10]. Dynamic programming tools decompose an original task into a several sub tasks, i.e. split with time, which can be solved separately. This approach may be easily used for a multi-constraint system and is very robust. However, it requires a careful selection of the initial data that will be analyzed and may be very difficult to extend to large river systems.

Other approaches are based on explicit stochastic optimization. In all cases optimization is performed without knowledge of the forecast. The modelling may be done using stochastic linear programming, stochastic dynamic programming, and stochastic optimal control. Such approaches require high computational capacities and thus may be hardly applied for large river basins. The last one, on the contrary, may be easily applied for large scales, however does not provide high precision in calculations. Another stochastic optimization approach is multiobjective optimization model. It allows setting many simultaneous constraints with subjective weights (relative magnitude of importance). In [11] this method is analyzed with four objectives: maximize energy production, improve energy production quality, minimize water discharges for water supply and maximize reliability of water supply.

There are known alternative approaches that were developed in 1970-s, but were widely used only in the beginning of 2000. These are, i.e. methods of linear programming and in particular adjustable robust optimization method that was developed as an alternative to the stochastic programming. In [12] such method was used for long-term planning of water usage in the Netherlands under unpredictable change of water level in the ocean. However, in this research authors admit high complexity of the calculations and its low applicability to the real life. In [13] a combination of Monte Carlo simulation with linear modelling is described to optimize power production at large power plants. However, it is obvious that usage of such technique requires knowledge of linear of linearizable mathematical model, which is impossible in practice. In [14] a control method for dynamic system of a hydropower plants cascade with a multi criteria solution. The solution is visualized as a 2D Pareto surface [15].

Today there are widely used mathematical modelling approaches based on the evolutionary self-organization algorithms [16]. So, in [17] an evolutionary model is described for power production control over the hydro power plants. In [18] another evolutionary algorithm is described that aims providing required water supply for water consumers during dry seasons.

One of the most used evolutionary algorithms is the genetic algorithm [19]. In [20] a simple and reliable approach is described for short-term forecasting and management that minimizes difference between power production and power consumption. In [21] usage of genetic algorithms results in development of water management plan for maximization of power production and water usage for irrigation. In [22] genetic algorithms were used for modelling of a simple cascade of three water basins with power production maximization. Another example of a combination of genetic algorithms with linear modelling for optimal rule curves of a reservoir at the Nam Oon river in Thailand is described in [23]. In the research, both of the methods are used separately and later the results are compared to choose the best one. In [24] optimal irrigation water allocation using a genetic algorithm under various weather conditions is described for different weather conditions that are described by probability precipitation, evaporation and water inflow to the reservoir. For each of the weather conditions there were two medels developed: model for complete irrigation and for deficite irrigation. The objective function is a profit maximization from harvesting disregarding weather conditions. In [25] an optimization method with genetic algorithms is described for a cascade of two reservoirs in Malaysia. The objective function optimizes power production and minimizes flood occurrence. The model is designed for a yearly modelling. The model may be considered as the most sophisticated and advanced among others, however it may not be used for other river systems, as far as it was initially created for a system for two consequent reservoirs. In [26] a genetic algorithm optimization model is described for irrigation planning from a single reservoir or a cascade of two reservoirs. Operation was planned in a 10-days term. The objective was to minimize the difference between planned and produced energy. The paper shows that usage of genetic algorithms is much more convenient compared to the other methods. In [27, 28] it is also shown a high correlation of a cascade of reservoirs modelling with genetic algorithms with linear programming. The case studies are given in India, China and the US.

Therefore, it is obvious that nowadays genetic algorithms are used for simple modeling of a small cascades of reservoirs with consideration of up to 3 criteria with the objective function to decrease the operation costs, or increase the profit from power production or harvesting, or reduce possible flood occurrence.

The known mathematical models for river networks may be greatly classified in to two categories. The first one studies a single reservoir, while the second one – a river network consisting of several reservoirs.

The mathematical models for a single reservoir normally include many criteria for reservoir management. So, in [29] an economic model of water management on a single reservoir is described. The model uses many criteria, such as required water supply for rice harvesting, for fishing, climate changes, and change of water allocated for household needs. In [30] an explicit mathematical model of a reservoir is described. It considers various inflows and outflows, surface water usage, household usage and others. Water management is performed in order to maintain either required water level or required water usage.

However, such models are rarely used in practice, as far as they cannot be assembled in a cascade of reservoirs with consideration of the processes that occur in each of the reservoir due to low computing capacity. Meanwhile, they are common for a single reservoir systems and allow to obtain a reliable and universal data on water management.

For a multi-reservoir river systems, the number of criteria used is greatly decreased (normally to a single criterion). According to [31] there are three main groups of multi-reservoir water management models. The first one studies maximization of water level in the reservoir under considered limit of power production. The second one maximizes power production. And the third one optimizes water allocation among consumers in order to maximize profits from its usage.

So, this paper aims usage of dynamic models with self-organization algorithms that may be used for construction of forecasting models with minimum information available about the reservoir system. The model will be built for a serial-parallel reservoir system with a multi criteria objective function to maintain required power production while maintaining required water level at the reservoir and minimizing the flood occurrence. The developed model should be used in a decision support software.

3 Software for Complex River Management

As it was shown before the main outcome of the developed mathematical model is a software for river management. The software mat be used as a decision support tool or for automated or semi-automated water management over the river system. So, all existing software may be divided into appropriate two groups.

The first group includes such well known tools as MIKE 11 by DHI (or MIKE Hydro River), Pusola and others.

MIKE 11 – the most popular tool for 1D river modeling. It is developed by DHI Water and Environment institute for flood areas modelling, analysis of flood aftermaths, dam breaches, and seasonal flood forecasting over a complex river system.

Pusola is a developed package for Delphi programming language. It is used for segments of the river network. It is versatile and may be used for to solve different tasks with almost unlimited modelling applications for ant hydrological systems with any boundary conditions. However, modelling is performed at pure Delphi programming language and thus it can be hardly used in real cases.

There is also known ECOMAG tool developed in Russia. It includes mathematical modelling, geo-information system, database with properties of territories and convenient graphical interface. The model applied is the spatial distributed model of the hydrological cycle, water discharge and contaminants transfer in the river basin. However, the tool doesn't calculate water levels. Therefore, it has to be used with combination of another tool such as MIKE 11 [32].

The second class of the software is used for reservoir management. Likewise the first group, an absolute world leader of the software tools is MIKE BASIN by DHI Water and Environment institute [33]. The software allows to perform planning and management of water resources of a single or of a group of reservoirs in order to develop strategic water use plan and solve many different water-related tasks. However, the versatility of the software limits its possible applications. The main problem is the impossibility to develop the water system operation rule based on the data imported related to allowable water discharges and other boundary conditions. So, MIKE Basin solves only the direct task: develop water level increase and assess possible flood levels, having estimated water discharges and limitations. No doubts, that such approach may be used for river system management, but it will require a lot of computation resources.

Among others there should be considered modelling and optimization tools that were developed in universal modelling tools, such as PowerSim, STELLA, iThink, Exted, and Venisim [34, 35]. However, they have lower detalization of the described processes can be hardly scaled to the real river networks and have narrow scope of application. Nevertheless, this paper will use PowerSim in order to build the above mentioned model.

Modelling will be performed on the region of Valle d'Aosta in Italy.

4 Geomorphology of Valle d'Aosta

The region Valle d'Aosta accounts for an important share of hydropower network system in Italy, consisting of mountainous region situated in North Western Italy bordered by Switzerland and France. Around 20% of this area is less than 1500 m from main sea level [36]. In actual practice the region consist of glaciers and snowpack in winter determine the runoff regime characterized by minimum flow values in winter and maximum flows in spring and summer. The principal river of the valley is 100 km through the whole Region between Courmayeur (near the Mont Blanc) and Quincinetto, (near the Pont St Martin) the outlet of the valley. The ice covers 5.5% of the total area and a great number of lakes are located in Valle d'Aosta. Some of such lakes are artificial and are used for the regulation of hydropower production. The availability of water to be stored in a certain elevation provides favorable conditions for hydroelectric productions. In the year 2011 the hydroelectric network capacity of the region was about 900 MW with a power production of 2743 GWh per year [37].

The first step in the modeling was the determination of an appropriate schematization of network of flows and reservoirs called as a hydropower system network. Every hydropower plant is characterized by the minimum and maximum flow of water available for the turbines and by an energetic coefficient that represents the actual power produced in MW. For each reservoir the maximum height must be provided for finding the power produced in each plant. Finally, the maximum flow and the time of concentration must be inserted for the dam and rivers connecting each other. These data, together with the initial and final conditions for the reservoir capacities, complete the conceptual model of the system.

The elevation and availability of water leads the network of reservoirs to function effectively. The Fig. 2 shows the topology of the region with terrain reflects the need for controlling the flow buy a simulation and optimization preventing flood without affecting power production.

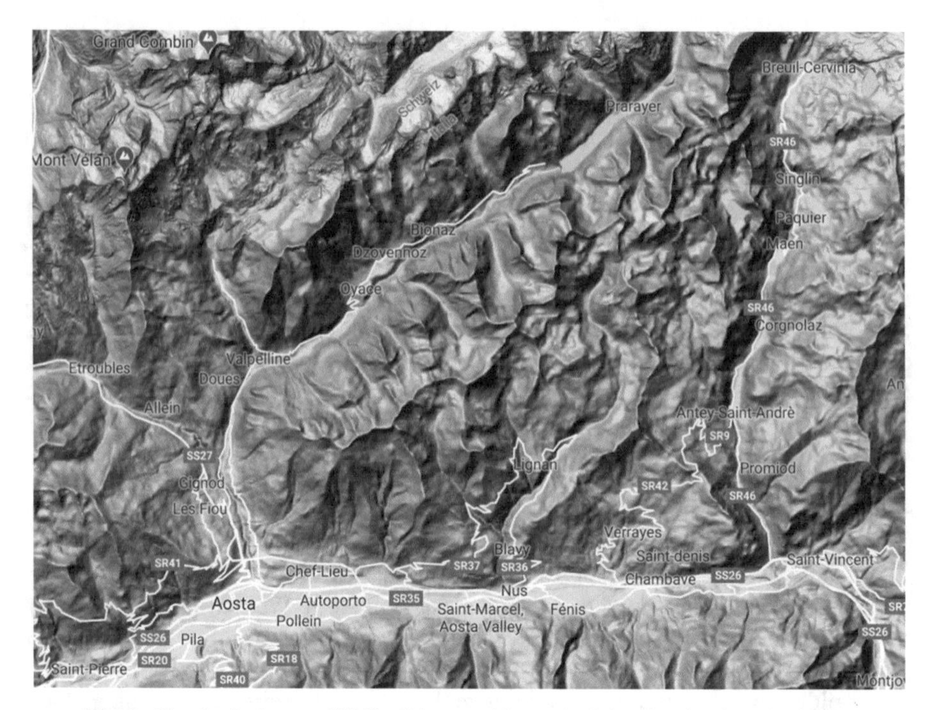

Fig. 2. Topological map of Valle d'Aosta with terrain (Map data by Google, 2019).

5 Modeling

Hydroelectric power plants may be located either in parallel to each other or in series. In this paper, it was selected a combination of two parallel reservoirs (A) and (B) and one installed in series after the parallel ones (C) in a way that the outlet water flow from reservoirs (A) and (B) falls into the third reservoir (C) (Fig. 3). Such a combination is considered to be one of the simplest for calculation but at the same time very commonly used.

In the very general case, the water balance model for each reservoir may be described as follows (see Fig. 4). The reservoir is supplied by the input stream from the upstream channels (Q_{in}). Additionally, rainwater (Q_r) is taken into account. The transition of underground water (Q_{und}) is also considered through the reservoir bed in positive (inside the reservoir) and in negative (to the ground) directions. Water may outgo from the reservoir due to evaporation (Q_{ev}), consumption for agricultural needs (Q_{irr}) and household maintenance (Q_{use}).

In the lower tail of the channel, the water is involved in the process of electricity generating (Q_{EP}) and may be discharged through the bypass (Q_{bp}) in order to increase the rate of emptying the reservoir. A sum of Q_{EP} and Q_{bp} in total gives the output flow (Q_{out}).

However, at this stage in the developed model evaporation is not considered because its calculation is complicated and will be added later. Therefore, the balance equation

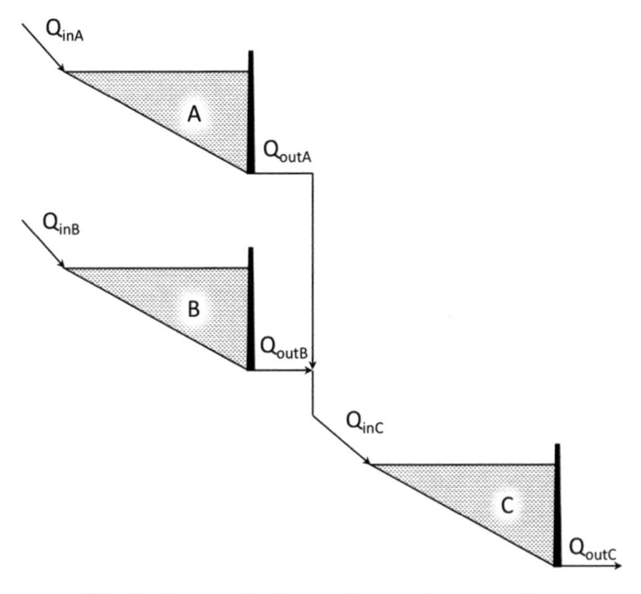

Fig. 3. A general scheme of reservoirs connection

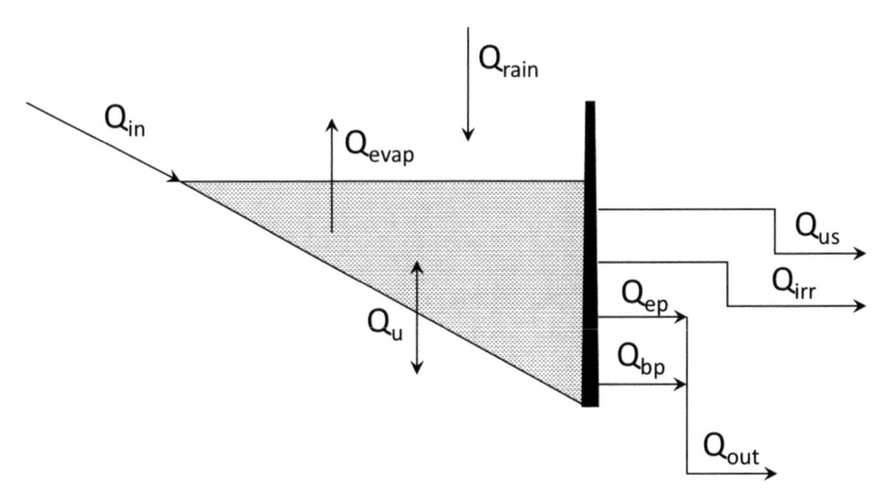

Fig. 4. Water flow model through a single reservoir

may be defined as follows [38]:

$$Q_{in} + Q_r \pm Q_{und} - Q_{irr} - Q_{use} - Q_{ep} - Q_{bp} = 0$$

where:

Q_{in} - incoming flow
Q_r - atmospheric precipitation

Q_{und} - underground water (could be positive or negative)
Q_{irr} - water, spent for irrigation
Q_{use} - water for household and industrial usage
Q_{ep} - discharge for energy production
Q_{bp} - discharge through bypass

The input data consists of the water amount required for household usage, irrigation and power production. A set of data including incoming flow due to rain precipitation, underground flows, and flow from the upstream reservoir represents a sum of all disturbances.

In order to perform an efficient water management, two strategies have been developed: an operating strategy and a planning strategy.

Operating strategy was developed with a simulation horizon 10 h, and moving horizon 3 h.

As inputs were set Q_{irr} - water, spent for irrigation, Q_{use} - water for household and industrial usage, Q_{ep} - discharge for energy production:

$$u_i = \begin{pmatrix} u_1 \\ u_2 \\ u_3 \\ u_4 \\ u_5 \\ u_6 \\ u_7 \\ u_8 \\ u_9 \end{pmatrix}$$

where:

u_1, u_2, u_3 - Q_{use} for 1st, 2nd and 3rd reservoirs, respectively;
u_4, u_5, u_6 Q_{irr} for 1st, 2nd and 3rd reservoirs, respectively;
u_7, u_8, u_9 - Q_{ep} for 1st, 2nd and 3rd reservoirs, respectively.

Also, a “pattern matrix” has to be entered, which consists of demand water supply:

$$\begin{pmatrix} u_{11} & \cdots & u_{1m} \\ \vdots & \ddots & \vdots \\ u_{n1} & \cdots & u_{nm} \end{pmatrix}$$

where

$j \in [1, m]$ and m - simulation horizon;
$i \in [1, n]$ and n - number of inputs (demands).

For each input its minimum $(u_{i_{min}})$ and maximum $(u_{i_{max}})$ values are defined, so that “pattern matrix” can change only within these limits:

$$u_i \in \left[u_{i_{min}}, u_{i_{max}}\right]$$

As disturbances were set Q_{in} (incoming flow), Q_r (atmospheric precipitation), Q_{und} (underground water):

$$d_i = \begin{pmatrix} d_1 \\ d_2 \\ d_3 \\ d_4 \\ d_5 \\ d_6 \\ d_7 \\ d_8 \\ d_9 \end{pmatrix}$$

where:

d_1, d_2, d_3 - Q_{in} for 1st, 2nd and 3rd reservoirs respectively;
d_4, d_5, d_6 - Q_r for 1st, 2nd and 3rd reservoirs respectively;
d_7, d_8, d_9 - Q_{und} for 1st, 2nd and 3rd reservoirs respectively.
So, the "disturbance matrix" may presented as follows:

$$\begin{pmatrix} d_{11} & \cdots & d_{1m} \\ \vdots & \ddots & \vdots \\ d_{k1} & \cdots & d_{km} \end{pmatrix}$$

where:

$j \in [1, m]$ and m - simulation horizon;
$i \in [1, k]$ and k - number of disturbances.
Incoming flow for 3rd reservoir is calculated as follows:

$$\mathrm{Q_{in_3}} = \mathrm{Q_{ep_1}} + \mathrm{Q_{ep_2}} + \mathrm{Q_{bp_1}} + \mathrm{Q_{bp_2}},$$

where:

$\mathrm{Q_{ep_1}}, \mathrm{Q_{ep_2}}$ – discharges for energy production in 1st and 2nd reservoirs respectively;
$\mathrm{Q_{bp_1}}, \mathrm{Q_{bp_2}}$ – bypass from 1st and 2nd reservoirs respectively.

For underground water flows a random matrix is created so that $\mathrm{Q_u} \in \left[\mathrm{Q_{u_{min}}}, \mathrm{Q_{u_{max}}}\right]$.

Discharge through bypass is a variable value, so it can be changed up to a certain maximum.

The elevations of reservoirs in region shown in network diagram of Valle d'Aosta (Fig. 2). The first step is to be making a conceptual model based on the given network. Each hydropower plant is characterized by a number of parameters such as minimum and maximum level of reservoir, flow rates, power produced. In addition to this describing the initial and final values required for each reservoir is also specified, completes the conceptual model.

We can consider a general case of hydroelectric power plant in order to understand the conceptual modelling because the region consists of 32 reservoirs mutually connected and we have to consider each dam separately before connecting it to a single network. Generally, the reservoir has three levels of capacity shown in Fig. 5, which are as follows: dead storage level (DSL, i.e., water level below which no electricity generation is possible), normal headwater level (NHL, i.e., accepted level in reservoir), surcharged reservoir level (SRL, i.e., water level aimed to store water during rainy season. In our case, we are considering only two basic levels the DSL and NHL. The reservoir will receive water from different sources like precipitation, ground water and water inflow from upstream reservoir.

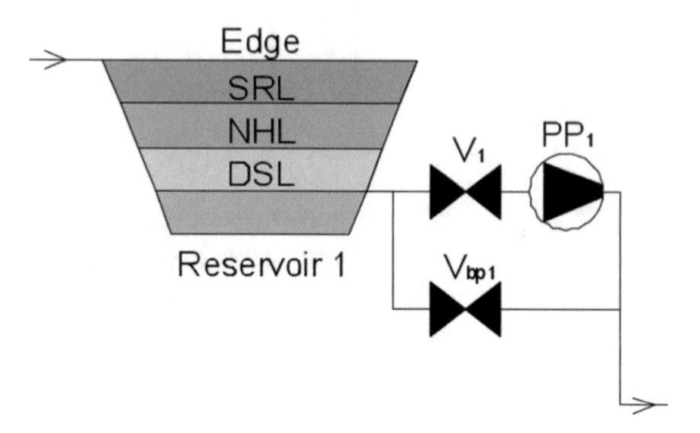

Fig. 5. A reservoir representing different levels of water based on constraints. SRL- Surcharged Reservoir Level, NHL- Normal Head water level and DSL - Dead Storage Level.

The management of hydroelectric power plant can be characterised by several constraints. Our main aim is to prevent from the overflow of water which results in the collapse of dam and this is prevented by maintaining a mass balance on each node and at each time interval.

So, in general and depending on boundary conditions we have to account for:

1. performance criteria – water level above the DSL the electricity generation range and below the SRL preventing from flood and collapse;
2. optimal mode of operation for maintaining the NHL.
3. production of rated power.

The software can be useful to compare different scenarios arising, for example, by a change of climatic conditions, or by different management politics. In this work, analysed scenario of water availability and simulation is done for a period of one year with an interval of one hour. The computer model begins with defining rated capacity, rated flow rate, rated power and head at which the plant works. In order to fulfil the boundary conditions an equation governing the flow rate has to be input which will keep the capacity of the reservoir within the DSL. Also, a mass balance is created by adding or exiting a specified flow rate according to our rated value. This constraint will limit the capacity within the DSL i.e. the minimum level will be returned to zero when capacity of reservoir reaches the minimum limit of dam.

The reservoir Beauregard is specifically analysed (Fig. 6). In the above equation we can find the dependence of minimum limit on flow rate and energy production based on flow rate, head and efficiency. All the variables are treated as constants because in our case they are used defined. A 10% of capacity is made for all the reservoirs and efficiency is considered to be 81%. But from the network we can clearly see that the power production cannot be maintained to the optimum value i.e. rated value because the capacity is decreasing with respect to time. So, in order to solve this here we are considering a mass balance on each power plant based on the flow rate at which it works.

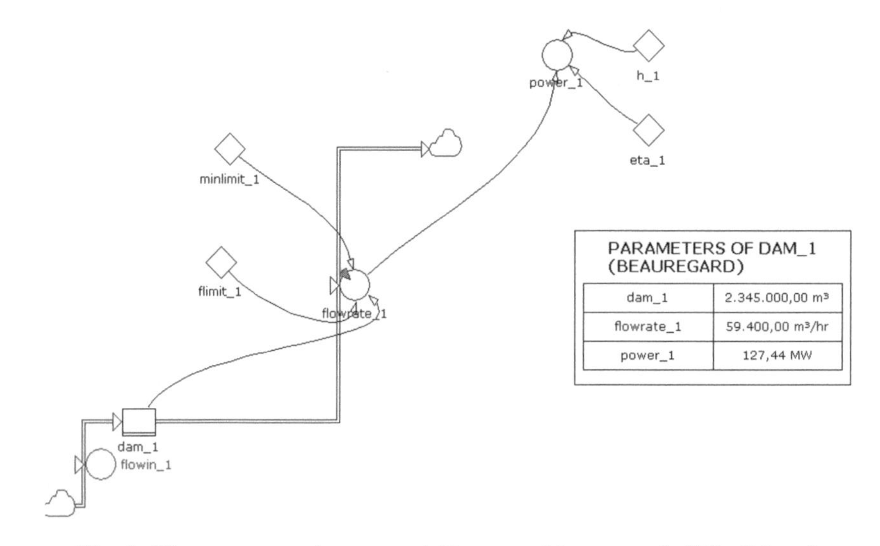

Fig. 6. Water movement in a reservoir Beauregard (a segment in Valle d'Aosta)

Similarly, all other 31 reservoirs are modelled and balanced. A segment of 4 out of 31 reservoirs is shown in Fig. 7. This mass balance can be assumed to be rainfall or from other resources which we took as a general case as flow-in our work, it may also from the exit of turbine to the next reservoir. Wide arrow indicates the direction of water course and thin arrow indicates the logical and structural relationship between the operators and flow chart elements. Next step is to cascade all this separately modelled power plant into a single network. This is accomplished by calling each separately modelled plant to a single network by using slice variable tool in Powersim. Subsequently we can simulate the obtained network for a period of one year from 2018 to 2019 with a time interval of one hour. Thus, the model is verified after one run of simulation. On interconnection some reservoirs will be affected with overflow or exceeding the SRL. This is slashed by creating a by-pass from the reservoir to the water channel.

6 Results

The proposed case study of Valle d'Aosta was modelled and successfully simulated in Powersim. It results in the creation of a real multi cascade river network and ensured

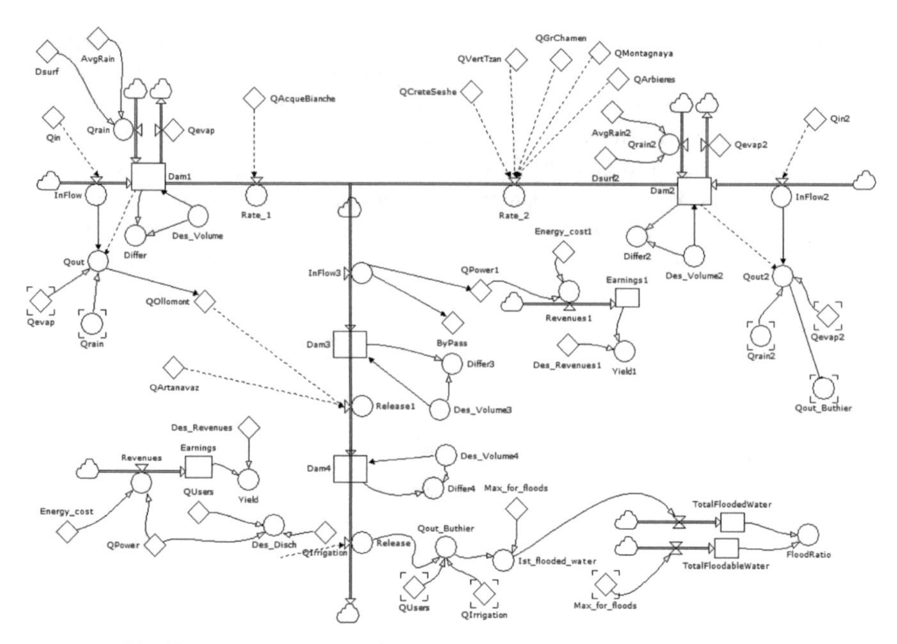

Fig. 7. Representation of whole coupled flow network in Valle d'Aosta

the water release from each dam based on the considered constraints. The major inputs shall include the following:

1. reservoir DHL and NHL levels
2. reservoir storage capacity
3. rated flow rate from reservoir
4. head of reservoir
5. efficiency of power plant.

After entering all the required data and initial conditions we can simulate the model by clicking a suitable button on the relevant control panel. The results are plotted in the Fig. 8. Graphical representation of simulation for a period of one year and we can see that the capacity of each reservoir is in its rated value i.e. it is maintained constant for each run and hence the modelled is verified and a balanced flow rate is obtained. The overall annual production is found to be 265.355 GW.

We can generate a dependence of power produced from each reservoir based on its capacity and it is plotted in excel by importing the simulated results into it shown in Fig. 8.

The Fig. 9 represents results of modelling of the Beauregard reservoir water level change with suggested modelling approach and without. The curve (1) on the Fig. 9 stands for change of water level in the reservoir in case of the existing modelling approach. It may be seen that high water level rise will be forecasted under set inflow forecast. However, introducing the suggested modelling approach will result in curve

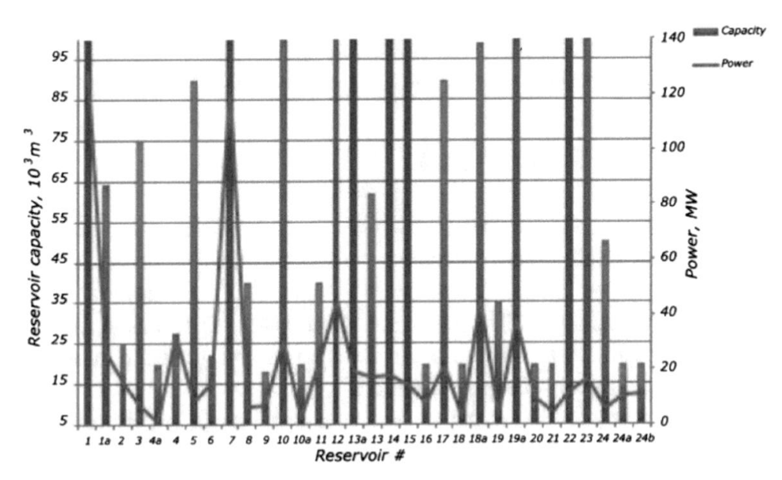

Fig. 8. Showing capacity and power produced of reservoirs in Valle d'Aosta

(2), which is even enough to prevent high water level rise at the reservoir and all other reservoirs also.

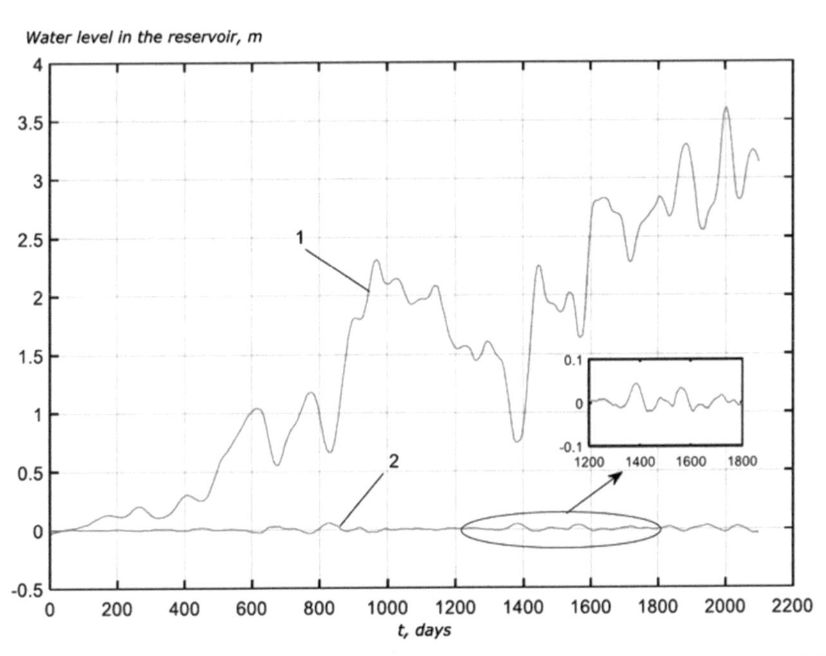

Fig. 9. Modelling of water level change at Beauregard reservoir with the suggested modelling approach (2) and with the existing modelling approach (1)

The model of the considered river network may be used to optimize a number of decisions concerning the management of the plant on the overall. For example, a simple goal may be that of maximizing the hydroelectric generation under the satisfaction of safety conditions. Thus, optimization cannot be separated by a careful risk analysis.

7 Conclusion and Future Work

The simulation performed gives maximum power which can be produced annually without uncertainties. In actual practice this may lead to uncontrolled distribution of water causing flood threatening life and investments. Balance flows of incoming flow, evaporation, and underground aquifers flows, bypass, reservoir, irrigation and domestic uses are taken into account. Each of these contributions can be represented with empirical formulae. In case of excessive accumulation the bypass which acts preventing structural failures. Based on such a modeling framework, optimization can be fruitfully applied for an effective management of the reservoir operations. Typical goals may be the following:

1. maintaining sufficient head leaves in reservoir to avoid the uncertainty of failure and flood due to overflow over spillways;
2. excluding the electricity consumption and storage, assumed complete consumption of produced power without wastage;
3. reducing the risk of flood by limiting the discharge flow of reservoirs.

By varying at different levels of consumption of water resources downstream such as domestic, agricultural, and electricity generation needs results effectively water management. Localized storage in agricultural fields can be implemented as a secondary means.

A sensitivity analysis can be done after an optimization or a risk analysis by implementing an actual uncertainty with highest probabilities an optimized reservoir control is possible.

Among the various approaches for water flow optimization, here we focus on genetic algorithms. Though they are not well developed yet and require vast monitoring set, high computing capacity, and help of an “expert” to “educate” their application, genetic algorithms appear to be appropriate in our context for various reasons. First of all, they do not require additional information such as the derivative of the cost or high-order derivatives. Thus, their application is straightforward inside in simulation tool like Powersim and, in principle, they allow to escape from local minima. By contrast, no guarantee to the convergence to the global point of optimum is ensured.

Modeling of hydroelectric systems with economic objectives are mostly difficult, it requires understanding the complex relationship between load market system and social safety. The model developed provide basis for an operational policy through numerous runs of simulation model throughout one year. This work is a way to develop powerful and transparent models to address hydroelectric generation systems of long term planning that are well-suited to being optimized. Optimization may aim at efficiently managing the reservoir operations to maximize the income from the power market and keep the river network far from flood. The key success of any optimization problem is its effective implementation based on the system features by using suitable mathematical model and proper algorithms. In the management of hydroelectric generation system, there are lots of uncertain and complex information. Future research should explore the use of uncertain information in system dynamics simulation.

References

1. Khosravi, K., et al.: A comparative assessment of flood susceptibility modeling using multi-criteria decision-making analysis and machine learning methods. J. Hydrol. **573**, 311–323 (2019). https://doi.org/10.1016/j.jhydrol.2019.03.073
2. Reuther, A.U., Herget, J., Ivy-Ochs, S., Borodavko, P., Kubik, P.W., Heine, K.: Constraining the timing of the most recent cataclysmic flood event from ice-dammed lakes in the Russian Altai Mountains, Siberia, using cosmogenic in situ 10Be. Geology **34**(11), 913–916 (2006)
3. Melillo, J.M., Richmond, T., Yohe, G.W. (eds.): Climate Change Impacts in the United States: The Third National Climate Assessment. U.S. Global Change Research Program, Washington, D.C. (2014)
4. Chen, H.H., Wang, L.: Hydropower simulation: an overview. Waterpower **87**, 803–812 (1987)
5. Jonkman, S.N., Kelman, I.: An analysis of the causes and circumstances of flood disaster deaths. Disasters **29**(1), 75–97 (2005). https://doi.org/10.1111/j.0361-3666.2005.00275.x
6. Kind, J.M.: Economically efficient flood protection standards for the Netherlands. J. Flood Risk Manage. **7**(2), 103–117 (2014)
7. Zhu, G., et al.: A new moving strategy for the sequential Monte Carlo approach in optimizing the hydrological model parameters. Adv. Water Resour. **114**, 164–179 (2018). https://doi.org/10.1016/j.advwatres.2018.02.007. ISSN 0309-1708
8. Hiew, K.: Optimization algorithms for large scale multi-reservoir hydropower systems.' Ph.D. dissertation, Dept. of Civil Engineering, Colorado State Univ., Ft. Collins, Colo (1987).
9. Lund, J., Ferreira, I.: Operating rule optimization for Missouri River reservoir system. J. Water Resour. Plan. Manage. **122**(4), 287–295 (1996)
10. Young, G.: Finding reservoir operating rules. J. Hydraul. Div. Am. Soc. Civ. Eng. **93**(6), 297–321 (1967)
11. Ko, S.-K., Fontane, D., Labadie, J.: Multiobjective optimization of reservoir systems operations. Water Resour. Bull. **28**(1), 111–127 (1992)
12. Postek, K., den Hertog, D., Kind, J., Pustjens, C.: Adjustable robust strategies for flood protection. Omega **82**, 142–154 (2018). https://doi.org/10.1016/j.omega.2017.12.009. ISSN 0305-0483
13. Piekutowski, M.R., Litwinowicz, T., Frowd, R.: Optimal short-term scheduling for a large-scale cascaded hydro system. In: Conference Proceedings Power Industry Computer Application Conference, pp. 292–298. IEEE (1993)
14. Zhu, F., Zhong, P.A., Xu, B., Wu, Y.N., Zhang, Y.: A multi-criteria decision-making model dealing with correlation among criteria for reservoir flood control operation. J. Hydroinformatics **18**(3), 531–543 (2016). https://doi.org/10.1007/s00477-016-1253-3
15. Karpenko, A.P., Moor, D.A., Mukhlisullina, D.T.: Multicriteria optimization based on neural network, fuzzy and neuro-fuzzy approximation of decision maker's utility function. Opt. Mem. Neural Netw. **21**(1), 1–10 (2012)
16. Deppermann, A., et al.: Increasing crop production in Russia and Ukraine—regional and global impacts from intensification and recultivation. Environ. Res. Lett. **13**(2), 025008 (2018). https://doi.org/10.1088/1748-9326/aaa4a4
17. Davidsen, C., et al.: Hydroeconomic optimization of reservoir management under downstream water quality constraints. J. Hydrol. **529**(3), 1679–1689 (2015). https://doi.org/10.1016/j.jhydrol.2015.08.018. ISSN 0022-1694

18. Al-Jawad, J.Y., Tanyimboh, T.T.: Reservoir operation using a robust evolutionary optimization algorithm. J. Environ. Manage. **197**, 275–286 (2017). https://doi.org/10.1016/j.jenvman.2017.03.081. ISSN 0301-4797
19. Andrey, P., Konstantin, N.: Intelligence control systems: contemporary problems in theory and implementation. Inf. Technol. Appl **2**, 72–78 (2012). Paneuropska vysoka skola
20. Lakshminarasimman, L., Subramanian, S.: Short-term scheduling of hydrothermal power system with cascaded reservoirs by using modified differential evolution. IEE Proc. Gener. Transm. Distrib. **153**(6), 693–700 (2006)
21. Devisree, M.V., Nowshaja, P.T.: Optimisation of reservoir operations using genetic algorithms. Int. J. Sci. Eng. Res. **5**(7), 340–344 (2014)
22. Hınçal, O., Altan-Sakarya, A.B., Ger, A.M.: Optimization of multireservoir systems by genetic algorithm. Water Resour. Manage. **25**(5), 1465–1487 (2011). https://doi.org/10.1007/s11269-010-9755-0
23. Suiadee, W., Tingsanchali, T.: A combined simulation–genetic algorithm optimization model for optimal rule curves of a reservoir: a case study of the Nam Oon Irrigation Project, Thailand. Hydrol. Process. Int. J. **21**(23), 3211–3225 (2007)
24. Sadati, S., Speelman, S., Sabouhi, M., Gitizadeh, M., Ghahraman, B.: Optimal irrigation water allocation using a genetic algorithm under various weather conditions. Water **6**(10), 3068–3084 (2014). https://doi.org/10.3390/w6103068
25. Fahimdanesh, M., Bahrami, M.E.: Evaluation of physicochemical properties of Iranian mango seed kernel oil. Int. Conf. Nutr. Food Sci. **53**(9), 44–49 (2013)
26. Asfaw, T.D., Saiedi, S.: Optimal short-term cascade reservoirs operation using genetic algorithm. Asian J. Appl. Sci. **4**(3), 297–305 (2011)
27. Raju, K.S., Kumar, D.N.: Irrigation planning using genetic algorithms. Water Resour. Manage. **18**(2), 163–176 (2004)
28. Liu, D., Chen, X., Lou, Z.: A model for the optimal allocation of water resources in a saltwater intrusion area: a case study in Pearl River Delta in China. Water Resour. Manage **24**(1), 63 (2010)
29. Tran, L.D., Schilizzi, S., Chalak, M., Kingwell, R.: Optimizing competitive uses of water for irrigation and fisheries. Agric. Water Manage. **101**(1), 42–51 (2011)
30. Belaineh, G., Peralta, R.C., Hughes, T.C.: Simulation/Optimization Modeling for Water Resources Management. J. Water Resour. Plann. Manage. **125**(3), 154–161 (1999)
31. Manos, B., Bournaris, T., Silleos, N., Antonopoulos, V., Papathanasiou, J.: A decision support system approach for rivers monitoring and sustainable management. Environ. Monit. Assess. **96**(1–3), 85–98 (2004)
32. Marks, D., Domingo, J., Susong, D., Link, T., Garen, D.: A spatially distributed energy balance snowmelt model for application in mountain basins. Hydrol. Process. **13**(12–13), 1935–1959 (1999)
33. Jaiswal, R.K., Ghosh, N.C., Guru, P.: MIKE BASIN based decision support tool for water sharing and irrigation management in Rangawan command of India. Adv. Agric. **2014**, 1–10 (2014). https://doi.org/10.1155/2014/924948
34. Ivanova, O., Neusipin, K., Ivanov, M., Schenone, M., Damiani, L., Revetria, R.: Optimization model of a tandem water reservoir system management. In: Proceedings of The World Congress on Engineering and Computer Science. LNCS, 25–27 October 2017, San Francisco, USA, pp. 893–899 (2017)
35. Revetria, R., Damiani, L., Ivanov, M., Ivanova, O.: An hybrid simulator for managing hydraulic structures operational modes to ensure the safety of territories with complex river basin from flooding. In: Proceedings of the 2017 Winter Simulation Conference, p. 220. IEEE Press (2017). Proceedings is free access is available at https://ieeexplore.ieee.org/
36. Berruto, G.: Una Valle d'Aosta, tante Valli d'Aosta? Considerazioni sulle dimensioni del plurilinguismo in una comunità regionale, pp. 44–53. FEC (2003)

37. Tibone, C., Agnesod, G., Cappio Borlino, M., Tartin, C., Crea, D., Berlier, F.: Noise model application to small hydroelectrical power plants impact evaluation in the Aosta Valley territory. Radiat. Prot. Dosimetry. **137**(3–4), 271–274 (2009)
38. Bittumon, K.B., Ivanov, M., Ivanova, O., Revetria, R., Sunjo, K.V.: Modeling for the Safe Management of Complex River Basins. In: Proceedings of the World Congress on Engineering and Computer Science. LNECS, 23–25 October 2018, San Francisco, USA, pp. 611–617 (2018)

A Logic Programming Solution to the Water Jug Puzzle

Feng-Jen Yang(✉)

Florida Polytechnic University, Lakeland, FL 33805, USA
fyang@floridapoly.edu

Abstract. Logic programming is usually preferred in deducing solutions for mental challenging problems such as solving puzzles or game playing in which the domain knowledge is encoded as production rules and the initial settings or inputs are represented as facts. By using the built-in inference engine that comes with the programming language itself, the solution to a given puzzle can be deduced and displayed. In this paper, a logic programming approach is demonstrated to solve the water jug puzzle.

Keywords: Forward chaining inference · Logical programming · Puzzle · Rule-based programming · State space · Search tree

1 Introduction

Logic programming is originated from formal logic with a special emphasis on the first-order predicate [1,2]. This programming paradigm is well known for its capability to emulate human's inference steps that is commonly held mentally. In real life human beings usually hold problem specific information in their short-term memory, but applying the knowledge in their long-term memory to deduce possible solutions. This manner of problem solving involves thee mechanisms, namely, the short-term memory, long-term memory, and deducing capability. In the digital world, these three mechanisms are realized as three typical components that forms a logical programming solution. In a logical program, a fact-base is used to hold problem specific information, a knowledge-base is used to hold domain knowledge, and a built-in inference engine is used to carry out the deductions toward the final solution by using the contents within the fact-base and knowledge-base. When a program is written, it is a programmer's responsibility to discover and encode the contents in the fact-base and knowledge-base. Since the inference engine is typically a built-in component provided by the programming language itself, it is waived from programming and not shown in a program.

Most of the real-life problems come with their particular domains. As an effort of analyzing and understanding a given problem, usually the problem domain can be visually represented as a semantic net. Based on the semantic

S.-I. Ao et al. (Eds.): WCECS 2018, *Transactions on Engineering Technologies*, pp. 66–73, 2020.
https://doi.org/10.1007/978-981-15-6848-0_6

net the programmer can go on to choose a suitable format to represent problem states at different point in time and encode knowledge rules that will cause state transitions within the problem domain [3]. In this manner, the entire problem domain can be viewed as a state space, also known as a search space, that consists of all possible states and all possible transitions among states. Within this state space, any path that starts from the initial state to a goal state is representing a solution. The series of operations that are performed along the path are representing the deducing steps of solving a problem. To prevent loops that might occur within a search space, any operation that will cause a transition from the current state to a previously visited state is excluded from the search space. In this paper, this problem-solving approach is illustrated to solve the water jug puzzle in CLIPS programming language.

2 The Water Jug Puzzle

The water jugs puzzle, also known as the water pouring problem, once was a popular civilian game in the medieval age. At that time people used to play this mathematical game by using a fixed number of water jugs that can hold different integral units of water volumes. A special constraint in this puzzle is that there is no measuring mark on the jugs so that a player has to rely mental planning and computations to have the expected amount of water in the designated jug . By filling up jugs from other jugs or emptying jugs into other jugs in a proper sequence, they found that they could end the game with having the exact amount of water they expected in advance in the jugs they dedicated. This kind of puzzle is a good illustration of problem solving by searching the state space to find a sequence of operations that can lead to a sequence of state transitions from the initial state to a goal state.

Through the ages, the water jug puzzle has evolved into several variations. For example, the game can start from for all jugs are full or some jugs are full and other jugs are empty. Sometimes, a tap and a sink are also provided so that besides allowing the player to fill up a jug by using the water in other jugs, it is also allowed to fill up a jug by using the tap. Similarly, besides allowing the player to empty the water in a jug into other jugs, it is also allowed to empty the water in a jug into the sink. When the tap and sink are provided, the game can also start from all jugs are empty [4].

In this paper, an instance of the water jug puzzle is selected for the purpose of demonstrating the logical programming approach towards puzzle solutions. In this instance, two water jugs are given in which one can hold up to 4 gallons of water and the other can hold up to 3 gallons of water. Both jugs have no measuring marks on them. The game rules are allowing a player to:

1. Fill up a jug from the tap.
2. Empty a jug into the sink.
3. Fill up a jug from the other jug.
4. Empty a jug into the other jug.

With these possible operations, the game starts from both jugs are full but at the end the game the player is required to fill the 4-gallon jug with exactly 2 gallons of water and leave the 3-gallon jug empty.

2.1 The Problem State Representation

To be able to construct a proper state space of the water jug puzzle, the first essential effort is to identify a suitable representation of each possible problem state. While solving the water jug puzzle, the only two variables that should be continuously tracked are the amount of water in each jug. This turns out that a very nature and intuitive way to represent a problem state is using an ordered pair of numbers to indicate the amount of water in the 4-gallon jug and the 3-gallon jug. So that the ordered pair (x, y) indicates that there are x gallons of water in the 4-gallon jug and y gallons of water in the 3-gallon jug. With this representation, this problem domain can be treated as a search space in which:

1. The initial state is (4, 3), because the game starts from both jugs are full.
2. The goal state is (2, 0) because the game ends with having exactly 2 gallons of water in the 4-gallon jug and leaves the 3-gallon jug empty.

2.2 The State Transitions

One of the major constraints on these puzzle is that there is no measuring mark on the jugs. Since there is no measuring mark to visualize the amount of water within each jug, it is better to have operations that will result in only integral units of water remained in the jugs. To this end, the only two categories of operations that are considered to be appropriate are filling up a jug and emptying a jug. Also, since a tap and a sink are also provided, a jug can either be filled up by using the tap or by using the water in the other jug. In a similar manner, a jug can either be emptied into the sink or into the other jug.

Overall, there are 8 possible operations that can be perform while solving this puzzle:

1. Fill up the 4-gallon jug from the tap, i.e., $(x, y) \rightarrow (4, y)$ where $x < 4$.
2. Fill up the 3-gallon jug from the tap, i.e., $(x, y) \rightarrow (x, 3)$ where $y < 3$.
3. Fill up the 4-gallon jug from the 3-gallon jug, i.e., $(x, y) \rightarrow (4, x+y-4)$ where $x > 0$ and $x+y \geq 4$.
4. Fill up the 3-gallon jug from the 4-gallon jug, i.e., $(x, y) \rightarrow (x+y-3, 3)$ where $y > 0$ and $x+y \geq 3$.
5. Empty the 4-gallon jug into the 3-gallon jug, i.e., $(x, y) \rightarrow (0, x+y)$ where $x > 0$ and $x+y \leq 3$.
6. Empty the 3-gallon jug into the 4-gallon jug, i.e. $(x, y) \rightarrow (x+y, 0)$ where $y > 0$ and $x+y \leq 4$.
7. Empty the 4-gallon jug into the sink, i.e., $(x, y) \rightarrow (0, y)$ where $x > 0$.
8. Empty the 3-gallon jug into the sin, i.e., $(x, y) \rightarrow (x, 0)$ where $y > 0$.

The operations 1 and 2 are derived from the 2 possible situations of filling up a jug from the tap. The operations 3 and 4 are derived from the 2 possible situations of filling up a jug from the other jug. The operations 5 and 6 are derived from the 2 possible situations of emptying a jug into the other jug. The operations 7 and 8 are derived from the 2 possible situations of emptying a jug into the sink.

3 The Implementation of Solution

By suing the CLIPS programming language, the solution can be programmed by asserting the initial state into the fact-base, and then encoding the 8 operations as inference rules in the knowledge-base. The work of inference are carried out the built-in inference engine of CLIPS [5]. The entire program is listed as follows [6]:

```
;Solving the Water Jug Puzzle in CLIPS
;Created by: Feng-Jen yang

;the fact-base

(deffacts the-initial-state
    (state 4 3))

;the rule-base

(defrule op1 "fill up the 4-gallon jug from the tap"
    (state ?x  ?y)
    (test (< ?x 4))
    (not (exists (state 4 ?y)))
    =>
    (assert (state 4 ?y))
    (assert (link  ?x ?y to 4 ?y)))

(defrule op2 "fill up the 3-gallon jug from the tap"
    (state ?x  ?y)
    (test (< ?y 3))
    (not (exists (state ?x 3)))
    =>
    (assert (state ?x 3))
    (assert (link  ?x ?y to ?x 3)))

(defrule op3 "fill up the 4-gallon jug from the 3-gallon jug"
    (state ?x ?y)
    (test (> ?x 0))
    (test (> (+ ?x ?y) 4))
    (not (exists (state 4 =(- (+ ?x ?y) 4))))
    =>
    (assert (state 4 =(- (+ ?x ?y) 4)))
    (assert (link ?x ?y to 4 =(- (+ ?x ?y) 4))))

(defrule op4 "fill up the 3-gallon jug from the 4-gallon jug"
    (state ?x ?y)
    (test (> ?y 0))
    (test (> (+ ?x ?y) 3))
    (not (exists (state =(- (+ ?x ?y) 3) 3)))
    =>
    (assert (state =(- (+ ?x ?y) 3) 3))
    (assert (link ?x ?y to =(- (+ ?x ?y) 3) 3)))

(defrule op5 "Empty the 4-gallon jug into the 3-gallon jug"
    (state ?x ?y)
    (test (> ?x 0))
```

```
    (test (< (+ ?x ?y) 3))
    (not (exists (state 0 =(+ ?x ?y))))
    =>
    (assert (state 0 =(+ ?x ?y)))
    (assert (link ?x ?y to 0 =(+ ?x ?y))))

(defrule op6 "Empty the 3-gallon jug into the 4-gallon jug"
    (state ?x ?y)
    (test (> ?y 0))
    (test (< (+ ?x ?y) 4))
    (not (exists (state =(+ ?x ?y) 0)))
    =>
    (assert (state =(+ ?x ?y) 0))
    (assert (link ?x ?y to =(+ ?x ?y) 0)))

(defrule op7 "Empty the 4-gallon jug into the sink"
    (state ?x ?y)
    (test (> ?x 0))
    (not (exists (state 0 ?y)))
    =>
    (assert (state 0 ?y))
    (assert (link ?x ?y to 0 ?y)))

(defrule op8 "Empty the 3-gallon jug into the sink"
    (state ?x ?y)
    (test (> ?y 0))
    (not (exists (state ?x 0)))
    =>
    (assert (state ?x 0))
    (assert (link ?x ?y to ?x 0)))

(defrule direct-path "constructing direct paths"
    (link ?x1 ?y1 to ?x2 ?y2)
    =>
    (assert (path ?x1 ?y1 to ?x2 ?y2 (str-cat "(" ?x1 ", " ?y1
    ") --> (" ?x2 ", " ?y2 ")"))))

(defrule indirect-path "constructing indirect paths"
    (path ?x1 ?y1 to ?x2 ?y2 ?route)
    (link ?x2 ?y2 to ?x3 ?y3)
    =>
    (assert (path ?x1 ?y1 to ?x3 ?y3 (str-cat ?route
    " --> (" ?x3 ", " ?y3 ")"))))

(defrule print-solutions "find and display solutions"
    (path 4 3 to 2 0 ?route)
    =>
    (printout t ?route crlf))
```

3.1 The Inferred Paths

While executing the program, the forward chaining inference engine is engaged in a series of inference cycles and trying to infer all possible paths. At the end, the following paths are inferred and the path from (4 3) to (2 0) is the solution to this instance of the water jug puzzle:

```
(path 4 3 to 0 3 "(4, 3) --> (0, 3)")
(path 0 3 to 3 0 "(0, 3) --> (3, 0)")
(path 4 3 to 3 0 "(4, 3) --> (0, 3) --> (3, 0)")
(path 3 0 to 4 0 "(3, 0) --> (4, 0)")
(path 0 3 to 4 0 "(0, 3) --> (3, 0) --> (4, 0)")
(path 4 3 to 4 0 "(4, 3) --> (0, 3) --> (3, 0) --> (4, 0)")
(path 4 0 to 0 0 "(4, 0) --> (0, 0)")
(path 3 0 to 0 0 "(3, 0) --> (4, 0) --> (0, 0)")
```

```
(path 0 3 to 0 0 "(0, 3) --> (3, 0) --> (4, 0) --> (0, 0)")
(path 4 3 to 0 0 "(4, 3) --> (0, 3) --> (3, 0) --> (4, 0) --> (0, 0)")
(path 3 0 to 3 3 "(3, 0) --> (3, 3)")
(path 0 3 to 3 3 "(0, 3) --> (3, 0) --> (3, 3)")
(path 4 3 to 3 3 "(4, 3) --> (0, 3) --> (3, 0) --> (3, 3)")
(path 3 3 to 4 2 "(3, 3) --> (4, 2)")
(path 3 0 to 4 2 "(3, 0) --> (3, 3) --> (4, 2)")
(path 0 3 to 4 2 "(0, 3) --> (3, 0) --> (3, 3) --> (4, 2)")
(path 4 3 to 4 2 "(4, 3) --> (0, 3) --> (3, 0) --> (3, 3) --> (4, 2)")
(path 4 2 to 0 2 "(4, 2) --> (0, 2)")
(path 3 3 to 0 2 "(3, 3) --> (4, 2) --> (0, 2)")
(path 3 0 to 0 2 "(3, 0) --> (3, 3) --> (4, 2) --> (0, 2)")
(path 0 3 to 0 2 "(0, 3) --> (3, 0) --> (3, 3) --> (4, 2) --> (0, 2)")
(path 4 3 to 0 2 "(4, 3) --> (0, 3) --> (3, 0) --> (3, 3) --> (4, 2)
--> (0, 2)")
(path 0 2 to 2 0 "(0, 2) --> (2, 0)")
(path 4 2 to 2 0 "(4, 2) --> (0, 2) --> (2, 0)")
(path 3 3 to 2 0 "(3, 3) --> (4, 2) --> (0, 2) --> (2, 0)")
(path 3 0 to 2 0 "(3, 0) --> (3, 3) --> (4, 2) --> (0, 2) --> (2, 0)")
(path 0 3 to 2 0 "(0, 3) --> (3, 0) --> (3, 3) --> (4, 2) --> (0, 2)
--> (2, 0)")
(path 4 3 to 2 0 "(4, 3) --> (0, 3) --> (3, 0) --> (3, 3) --> (4, 2)
--> (0, 2) --> (2, 0)")
(path 2 0 to 2 3 "(2, 0) --> (2, 3)")
(path 0 2 to 2 3 "(0, 2) --> (2, 0) --> (2, 3)")
(path 4 2 to 2 3 "(4, 2) --> (0, 2) --> (2, 0) --> (2, 3)")
(path 3 3 to 2 3 "(3, 3) --> (4, 2) --> (0, 2) --> (2, 0) --> (2, 3)")
(path 3 0 to 2 3 "(3, 0) --> (3, 3) --> (4, 2) --> (0, 2) --> (2, 0)
--> (2, 3)")
(path 0 3 to 2 3 "(0, 3) --> (3, 0) --> (3, 3) --> (4, 2) --> (0, 2)
--> (2, 0) --> (2, 3)")
(path 4 3 to 2 3 "(4, 3) --> (0, 3) --> (3, 0) --> (3, 3) --> (4, 2)
--> (0, 2) --> (2, 0) --> (2, 3)")
(path 2 3 to 4 1 "(2, 3) --> (4, 1)")
(path 2 0 to 4 1 "(2, 0) --> (2, 3) --> (4, 1)")
(path 0 2 to 4 1 "(0, 2) --> (2, 0) --> (2, 3) --> (4, 1)")
(path 4 2 to 4 1 "(4, 2) --> (0, 2) --> (2, 0) --> (2, 3) --> (4, 1)")
(path 3 3 to 4 1 "(3, 3) --> (4, 2) --> (0, 2) --> (2, 0) --> (2, 3)
--> (4, 1)")
(path 3 0 to 4 1 "(3, 0) --> (3, 3) --> (4, 2) --> (0, 2) --> (2, 0)
--> (2, 3) --> (4, 1)")
(path 0 3 to 4 1 "(0, 3) --> (3, 0) --> (3, 3) --> (4, 2) --> (0, 2)
--> (2, 0) --> (2, 3) --> (4, 1)")
(path 4 3 to 4 1 "(4, 3) --> (0, 3) --> (3, 0) --> (3, 3) --> (4, 2)
--> (0, 2) --> (2, 0) --> (2, 3) --> (4, 1)")
(path 4 1 to 0 1 "(4, 1) --> (0, 1)")
(path 2 3 to 0 1 "(2, 3) --> (4, 1) --> (0, 1)")
(path 2 0 to 0 1 "(2, 0) --> (2, 3) --> (4, 1) --> (0, 1)")
(path 0 2 to 0 1 "(0, 2) --> (2, 0) --> (2, 3) --> (4, 1) --> (0, 1)")
(path 4 2 to 0 1 "(4, 2) --> (0, 2) --> (2, 0) --> (2, 3) --> (4, 1)
--> (0, 1)")
(path 3 3 to 0 1 "(3, 3) --> (4, 2) --> (0, 2) --> (2, 0) --> (2, 3)
--> (4, 1) --> (0, 1)")
(path 3 0 to 0 1 "(3, 0) --> (3, 3) --> (4, 2) --> (0, 2) --> (2, 0)
--> (2, 3) --> (4, 1) --> (0, 1)")
(path 0 3 to 0 1 "(0, 3) --> (3, 0) --> (3, 3) --> (4, 2) --> (0, 2)
--> (2, 0) --> (2, 3) --> (4, 1) --> (0, 1)")
(path 4 3 to 0 1 "(4, 3) --> (0, 3) --> (3, 0) --> (3, 3) --> (4, 2)
--> (0, 2) --> (2, 0) --> (2, 3) --> (4, 1) --> (0, 1)")
(path 0 1 to 1 0 "(0, 1) --> (1, 0)")
(path 4 1 to 1 0 "(4, 1) --> (0, 1) --> (1, 0)")
(path 2 3 to 1 0 "(2, 3) --> (4, 1) --> (0, 1) --> (1, 0)")
(path 2 0 to 1 0 "(2, 0) --> (2, 3) --> (4, 1) --> (0, 1) --> (1, 0)")
(path 0 2 to 1 0 "(0, 2) --> (2, 0) --> (2, 3) --> (4, 1) --> (0, 1)
--> (1, 0)")
(path 4 2 to 1 0 "(4, 2) --> (0, 2) --> (2, 0) --> (2, 3) --> (4, 1)
--> (0, 1) --> (1, 0)")
(path 3 3 to 1 0 "(3, 3) --> (4, 2) --> (0, 2) --> (2, 0) --> (2, 3)
```

```
--> (4, 1) --> (0, 1) --> (1, 0)")
(path 3 0 to 1 0 "(3, 0) --> (3, 3) --> (4, 2) --> (0, 2) --> (2, 0)
--> (2, 3) --> (4, 1) --> (0, 1) --> (1, 0)")
(path 0 3 to 1 0 "(0, 3) --> (3, 0) --> (3, 3) --> (4, 2) --> (0, 2)
--> (2, 0) --> (2, 3) --> (4, 1) --> (0, 1) --> (1, 0)")
(path 4 3 to 1 0 "(4, 3) --> (0, 3) --> (3, 0) --> (3, 3) --> (4, 2)
--> (0, 2) --> (2, 0) --> (2, 3) --> (4, 1) --> (0, 1) --> (1, 0)")
(path 1 0 to 1 3 "(1, 0) --> (1, 3)")
(path 0 1 to 1 3 "(0, 1) --> (1, 0) --> (1, 3)")
(path 4 1 to 1 3 "(4, 1) --> (0, 1) --> (1, 0) --> (1, 3)")
(path 2 3 to 1 3 "(2, 3) --> (4, 1) --> (0, 1) --> (1, 0) --> (1, 3)")
(path 2 0 to 1 3 "(2, 0) --> (2, 3) --> (4, 1) --> (0, 1) --> (1, 0)
--> (1, 3)")
(path 0 2 to 1 3 "(0, 2) --> (2, 0) --> (2, 3) --> (4, 1) --> (0, 1)
--> (1, 0) --> (1, 3)")
(path 4 2 to 1 3 "(4, 2) --> (0, 2) --> (2, 0) --> (2, 3) --> (4, 1)
--> (0, 1) --> (1, 0) --> (1, 3)")
(path 3 3 to 1 3 "(3, 3) --> (4, 2) --> (0, 2) --> (2, 0) --> (2, 3)
--> (4, 1) --> (0, 1) --> (1, 0) --> (1, 3)")
(path 3 0 to 1 3 "(3, 0) --> (3, 3) --> (4, 2) --> (0, 2) --> (2, 0)
--> (2, 3) --> (4, 1) --> (0, 1) --> (1, 0) --> (1, 3)")
(path 0 3 to 1 3 "(0, 3) --> (3, 0) --> (3, 3) --> (4, 2) --> (0, 2)
--> (2, 0) --> (2, 3) --> (4, 1) --> (0, 1) --> (1, 0) --> (1, 3)")
(path 4 3 to 1 3 "(4, 3) --> (0, 3) --> (3, 0) --> (3, 3) --> (4, 2)
--> (0, 2) --> (2, 0) --> (2, 3) --> (4, 1) --> (0, 1) --> (1, 0)
--> (1, 3)")
```

3.2 The Solution Displayed as a Path

When the last inference rule is fired, all non-solution paths are screened out and the valid solution is identified and displayed as follows:

```
(4, 3) --> (0, 3) --> (3, 0) --> (3, 3) --> (4, 2) --> (0, 2) --> (2, 0)
```

4 Conclusion

Solving a puzzle is mentally challenging but a lot of fun. By nature most of the steps toward the final solution are intensively counting on inference. This intrinsic nature makes logic programming a good choice for deducing solutions to a puzzle. By taking advantages of the powerful built-in inference engine that usually comes with a programming is the logical paradigm such as CLIPS and PROLOG, the programming of soling a puzzle is much easier. Instead of writing tedious inference procedures by using Object-Oriented programming languages, Logical programming formulate human inference knowledge into production rules and count the built-in inference engine to carry out the inference process. Even though most of the college students are more familiar and comfortable with imperative and object-oriented paradigms of programming, logical programming paradigm is more suitable of solving AI problems. In this paper, a sample program in CLIPS programming language is illustrated to solve the water jug puzzle. Instead of aiming at inventing new theory or problem-solving method, this illustration is for the purpose of providing an additional problem-solving example to AI related studies.

References

1. Bratko, I.: Prolog Programming for Artificial Intelligence, 4th edn. Addison Wesley, NY, USA (2012)
2. Clocksin, W.F., Mellish, C.S.: Programming in Prolog. Springer, New York (2003)
3. Baral, C.: Knowledge Representation, Reasoning and Declarative Problem Solving. Cambridge University Press, Cambridge (2003)
4. Cowley, E.B.: Note on a linear diophantine equation, questions and discussions. Am. Math. Mon. **33**(7), 379–381 (1926)
5. Giarratano, J.C.: CLIPS 6.4 User's Guide (2018). http://clipsrules.sourceforge.net/documentation/
6. Yang, F.-J.: Solving the water jug puzzle in CLIPS. In: Proceedings of The World Congress on Engineering and Computer Science. LNECS, 23–25 Oct 2018, San Francisco, USA, pp. 564–567 (2018)

Diary Factory ProModel Modeling and Simulation, Layout Assessment and Improvement

Khaled M. Toffaha(✉) and Shi Dongyan

Collage of Mechanical Engineering, Harbin Engineering University, Harbin 150001, China
{Khaled.Toffaha,ShiDongyan}@hrbeu.edu.cn

Abstract. Recently, dairy farming has been transformed into an industrialized system. As a result, dairy industry has been endeavoring to develop their operations through the adopting of several improvement methods and techniques. The current paper aims at evaluating existing operations and processes at a Dairy Factory in the state of Palestine, and to improve the efficiency of its operations and thus obtain maximum productivity. The scope of work in this paper is to use simulation techniques to build a model that represents the processes at the factory, the model runs for 2000 h using (ProModel) also using Facility plant layout approach to evaluates current factory layout and suggests new improved one based on certain calculations.

Keywords: Diary factory · Corelap layout · Layout assessment and improvement · Process flow chart · Plant layout · ProModel simulation · Plant modeling and simulation · Productivity and benchmarking

1 Introduction

1.1 Introduction to the Dairy Factory

Dairy factory is one of the major productive charitable projects in the state of Palestine. It produces dairy products such as, Ultra heat temperature (UHT), Choco, cheese, condense milk, sour milk, sour milk up, yogurt, Shemaint and juice. At 2001, the factory started Ultra heat temperature milk production. This project was extracted to implementation site in order to help Palestinian farmers to utilize their farms and sell their own milk, also to provide dairy products to the Palestinian national economy. At the beginning, Tetra Park Company provided the factory with machines, prepare design, equipment and training to the technical cadres according to the latest Quality System.

1.2 Introduction to the Analysis

Methods engineering is a technique used to improve productivity and reduce costs in both direct and indirect operations of manufacturing and non-manufacturing business

S.-I. Ao et al. (Eds.): WCECS 2018, *Transactions on Engineering Technologies*, pp. 74–86, 2020.
https://doi.org/10.1007/978-981-15-6848-0_7

organizations. It can be defined as the systematic procedure for subjecting all direct and indirect operations. We used engineering methods to draw the process flow chart which is "a type of diagram that represents an algorithm, workflow or process, showing the steps as boxes of various kinds, and their order by connecting them with arrows, in this chart we draw each operation step for each manufacturing process, in order to analyze the steps and try to eliminate combine or improve it.

Plant Simulation is a computer application for modeling, simulating, analyzing, visualizing and optimizing production systems and processes, the flow of materials and logistic operations. Simulation can optimize material flow, resource utilization and logistics for all levels of plant planning from global production facilities, through local plants, to specific lines. The application allows comparing complex production alternatives, including the immanent process logic. Plant Simulation is used primarily to strategically plan layout, control logic and dimensions of large, complex production investments.

Plant layout design has become a fundamental basis of today's industrial plants which can influence parts of work efficiency. It is needed to appropriately plan and position employees, materials, machines, equipment, and other manufacturing supports and facilities to create the most effective plant layout. A plant layout study is an engineering study used to analyze different physical configurations for a manufacturing plant."

2 Methodology

2.1 Process Flow Chart

We have followed the flow chart principles: by defining each process in order to be diagrammed, discussing and deciding on the boundaries of each process: Where or when does, the process start, also where or when does it end, discussing and deciding the level of the details to be included in the diagram. Brainstorm the activities that take place, determining the process sequence, and arranging the activities in proper sequence.

2.2 Plant Simulation

In order to perform diary plant simulation, we used ProModel which is a discrete-event simulation technology that is used to plan, design and improve new or existing manufacturing, logistics and other operational systems. It empowers you to accurately represent real-world processes, including their inherent variability and interdependencies, in order to conduct predictive analysis on potential changes. Optimize your system around your key performance indicators. The ProModel Methodology:

Visualize: Create a dynamic, animated computer model of your business environment from CAD files, process or value stream maps, or Process Simulator models. Clearly see and understand current processes and policies in action.

Analyze: Brainstorm using the model to identify potential changes and develop scenarios to test improvements which will achieve business objectives. Run scenarios independently of each other and compare their results in the Output Viewer developed through the latest Microsoft® WPF technology.

Optimize: Immediately test the impact of changes on current and future operations, risk free, with predictive scenario comparisons. Determine optimal business performance with a high probability of meeting your business goals.

2.3 Plant Layout

The six tools and techniques used for layout planning/plant layout are as follows [2]: 1. Operation Process Chart: The manufacturing process is divided into separate operations with the help of the operation process chart. It shows the points at which materials are introduced into the process and the sequence of various operations and inspections other than material handling. 2. Flow Process Chart: This chart is a graphic representation of all the production activities occurring on the shop floor, which includes transportation, storage and delay. 3. Process Flow Diagrams: It is the diagram of building plan representing graphically the relative position of productive machinery storage space, gangways etc. and path followed by men or materials. All routes followed by different items are shown by joining symbols with straight lines. 4. Machine Data Cards: These cards give complete specification of each machine to be installed showing its capacity, space and other requirements, foundations methods of operation, maintenance and handling devices of machines etc. 5. Templates: After studying the flow process chart, process flow diagram and machine data cards, a floor plan is prepared by fixing the area occupied by each item (machine/equipment, benches, racks, material handling equipment etc.) to be erected in the shops. These templates are arranged in such a way so as to provide the best layout. This procedure makes the layout visual before actually drawn and is carefully examined. 6. Scale Models: It is an improvement over the template technique. In this tool, instead of templates, three-dimensional scale model is utilized. These models may be of wood plastic or metals. When these are used on a layout, series of additional information about the height and of the projected components of the machines are obtained.

3 Results

3.1 Process Flow Chart "Fig. 1"

3.2 Plant Simulation

90 readings of arrivals have been gathered and analyzed using ProModel [3] Stat. Fit. The distribution resulted was Exponential with mean 42.9 min. In addition, scatter plot, autocorrelation and runs tests have been conducted and indicate non-correlation in the data.

The following figures show the gathered data as well as ProModel Stat results (Fig. 2):

The following table illustrates the main distributions for each activity as well as its fitting plot as analyzed in Stat Fit (Table 1):

3.2.1 Model Building (ProModel)

3.2.1.1 Locations

All the needed locations were built and named, there are fifteen different locations in this model, and each location represents distinct process or resource. Furthermore, each location has assigned specific capacity. The following figure shows the locations as appear in ProModel (Fig. 3):

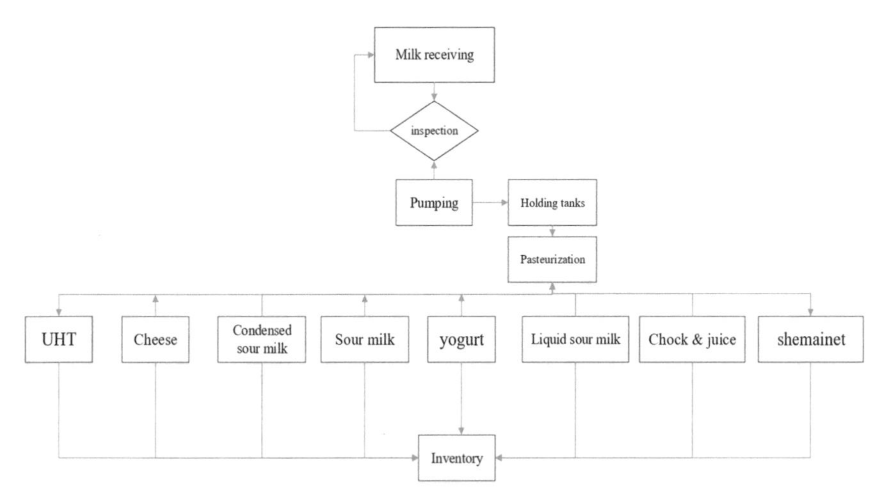

Fig. 1. Process flow chart

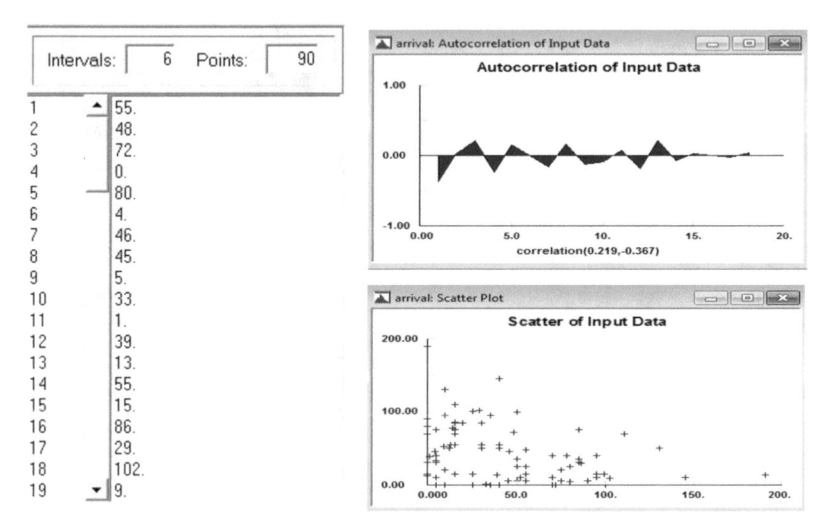

Fig. 2. Inter arrivals analysis

3.2.1.2 Entities

There is only one entity which is Milk, the milk arrives to the receiving area, route in different processes and produced in many forms. The following figure shows the entities as defined in ProModel (Fig. 4):

3.2.1.3 Arrivals

The Milk arrive to the factory from farmers in an exponential distribution and in an average quantity of 1000 L per farmer. The following figure shows the arrivals as defined in Pro Model (Fig. 5):

Table 1. Distribution fitting

Activity	Distribution	Distribution fit
Inter arrivals	Exponential (42.9)	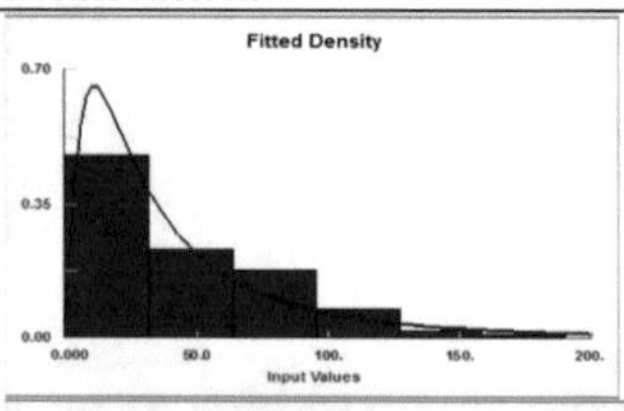
Inspection	Lognormal (3, 1.2)	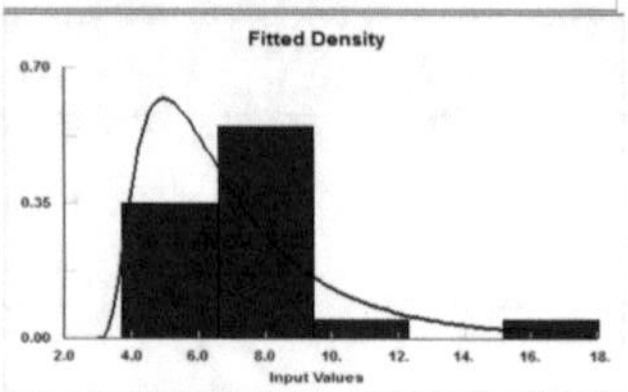
Pump	Lognormal (1, 1.26)	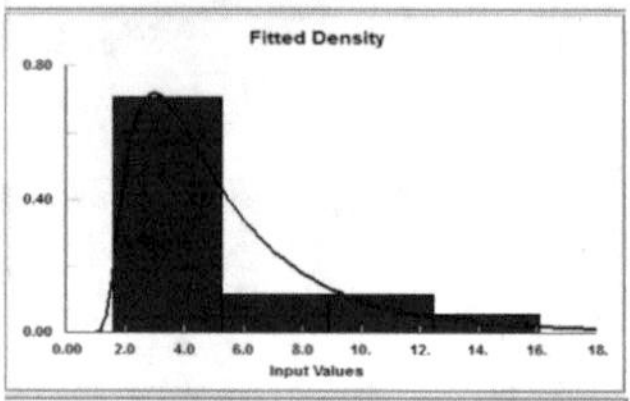
Milk Packaging	Lognormal (0.833,3.23)	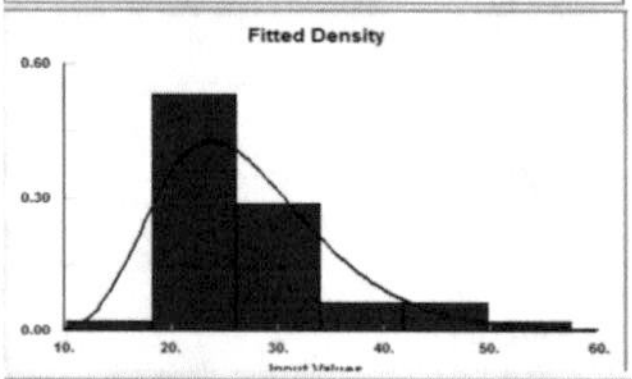
Sour Milk	Lognormal (13, 0.8)	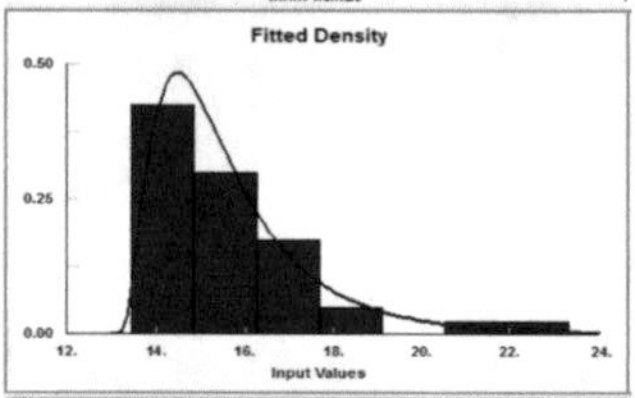
Yogurt	Lognormal (0.447,3.19)	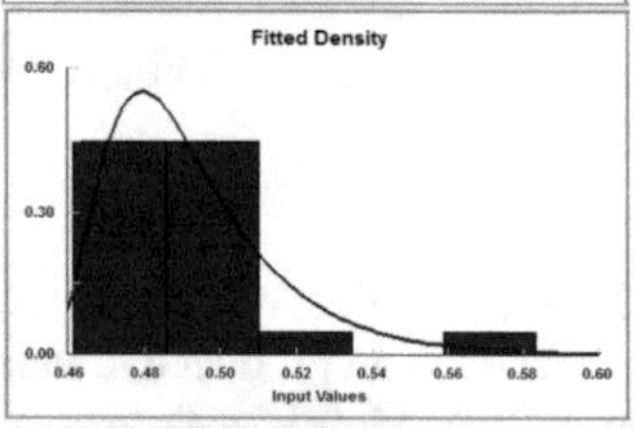
Choco	Lognormal (0.163,2.71)	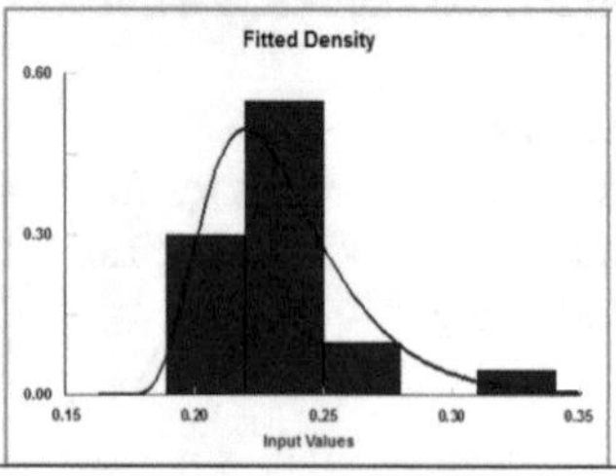

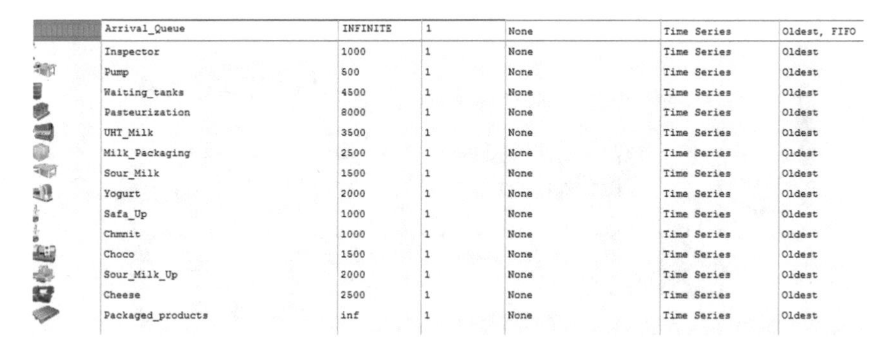

	Arrival_Queue	INFINITE	1	None	Time Series	Oldest, FIFO
	Inspector	1000	1	None	Time Series	Oldest
	Pump	500	1	None	Time Series	Oldest
	Waiting_tanks	4500	1	None	Time Series	Oldest
	Pasteurization	8000	1	None	Time Series	Oldest
	UHT_Milk	3500	1	None	Time Series	Oldest
	Milk_Packaging	2500	1	None	Time Series	Oldest
	Sour_Milk	1500	1	None	Time Series	Oldest
	Yogurt	2000	1	None	Time Series	Oldest
	Safa_Up	1000	1	None	Time Series	Oldest
	Chmnit	1000	1	None	Time Series	Oldest
	Choco	1500	1	None	Time Series	Oldest
	Sour_Milk_Up	2000	1	None	Time Series	Oldest
	Cheese	2500	1	None	Time Series	Oldest
	Packaged_products	inf	1	None	Time Series	Oldest

Fig. 3. Locations

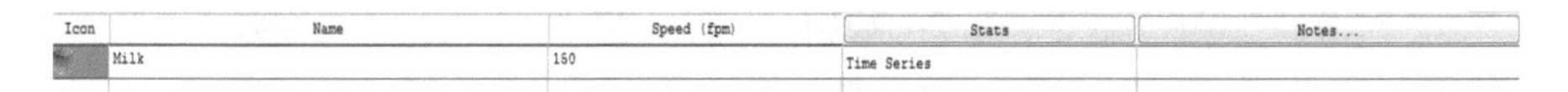

Icon	Name	Speed (fpm)	Stats	Notes...
	Milk	150	Time Series	

Fig. 4. Entities

Arrivals

Entity...	Location...	Qty Each...	First Time...	Occurrences	Frequency	Logic...	Disable
Milk	Arrival_Queue	1000		inf	E(42.9)		No

Fig. 5. Arrivals

3.2.1.4 Layout

The following figure shows the layout of the process as defined in ProModel, it shows the main locations and their approximate arrangements [4] (Fig. 6):

3.2.1.5 Processing

Processing were clearly defined and identified for each process; the following figure shows the processing as appears in ProModel (Fig. 7):

After building the model, it has run for 2000 h to obtain a reasonable output and observe the system at steady state values. The total produced milk during 1000 h time was 2,785,000 L with 25.91 min average time in system per liter. The following figure shows the sates for multiple capacity locations during 2000 h run time (Fig. 8):

The utilization for each location is shown in the following figure (Fig. 9):

The following figure shows the locations summary during the simulation run time (Fig. 10):

The following figure shows the entities summary during the simulation run time (Fig. 11):

3.3 Plant Layout

The aim of this tool is to assess and evaluate the current layout of the Dairy factory, also trying to build new layout for the dairy through implementing layout methods of evaluating and constructing. We've measured the areas and the distances of each department in the factory to get (Fig. 12)

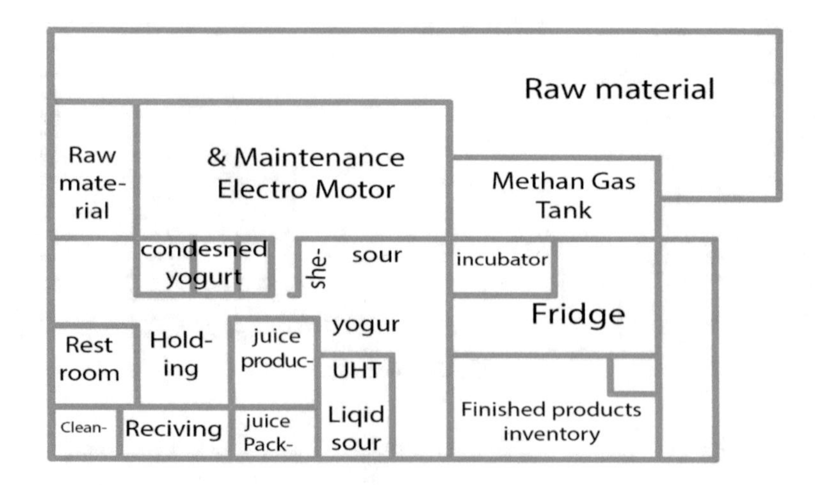

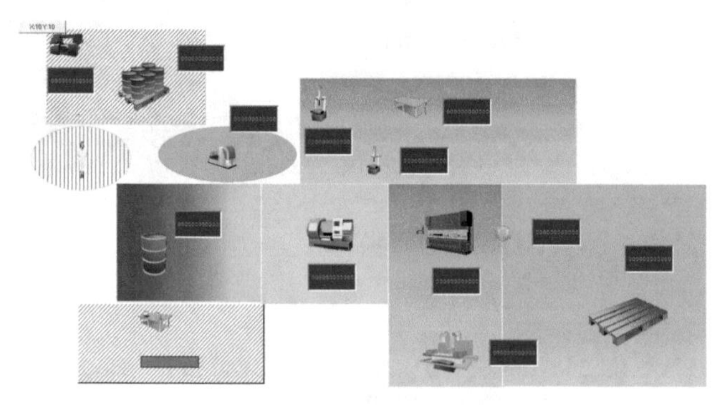

Fig. 6. Layout

Milk	Arrival_Queue	
Milk	Inspector	Wait L(3, 1.2)
Milk	Pump	Wait L(1, 1.26)
Milk	Waiting_tanks	
Milk	Pasteurization	Wait 19
Milk	UHT_Milk	Wait .0153
Milk	Milk_Packaging	Wait L(.833, 3.23)
Milk	Sour_Milk	Wait L(13, .8)
Milk	Yogurt	Wait L(.447, 3.19)
Milk	Safa_Up	Wait .13 sec
Milk	Chmnit	Wait 4.27 sec
Milk	Choco	Wait L(.163, 2.71)
Milk	Sour_Milk_Up	Wait .055 min
Milk	Cheese	Wait E(0.217)
Milk	Packaged_products	

Fig. 7. Processing

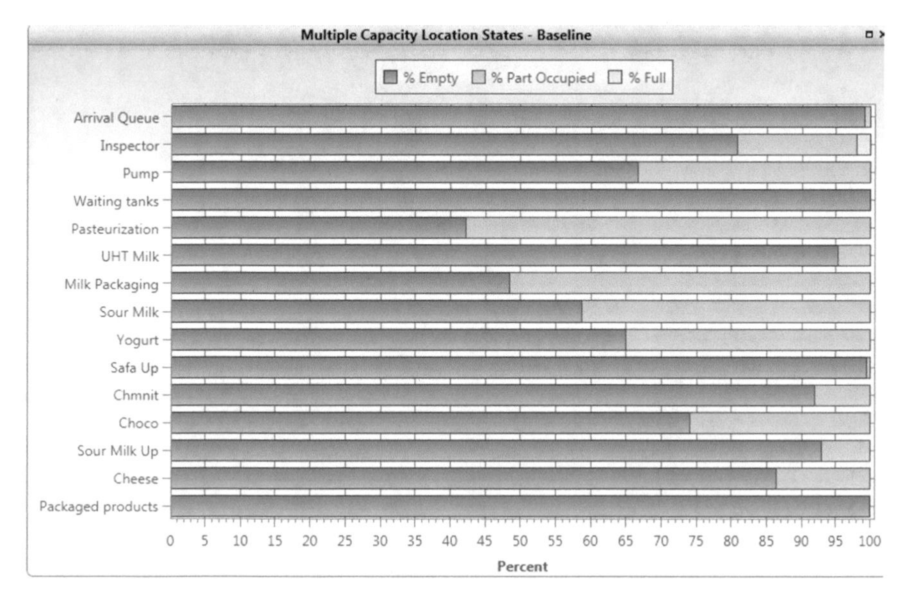

Fig. 8. Locations state

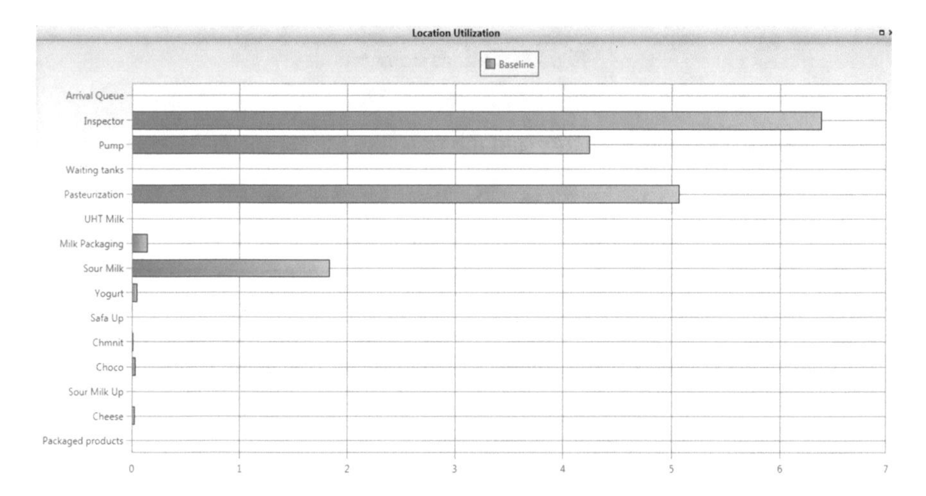

Fig. 9. Locations utilization

We've tried to follow the process of the Dairy in order to measure the areas and the distances, After the observation of the flow of operators between departments we constructed a table and found that the higher flow was between incubator and fridge departments. Due to the large amount of product that must be in the incubator for hours to take the good substratum. After that we evaluate the relationship chart between departments, we divided the largest number of flows by 5 and built this closeness relating ships ranges. We have only one absolutely important relationship between incubator and fridge departments and only two ordinary importance relationships between Safeno

Location Summary

Name	Scheduled Time (Hr)	Capacity	Total Entries	Average Time Per Entry (Min)	Average Contents	Maximum Contents	Current Contents	% Utilization
Arrival Queue	2,000.00	999,999.00	2,785,000.00	0.22	5.22	2,011.00	0.00	0.00
Inspector	2,000.00	1,000.00	2,785,000.00	3.00	69.62	1,000.00	0.00	6.96
Pump	2,000.00	500.00	2,785,000.00	1.00	23.23	378.00	0.00	4.65
Waiting tanks	2,000.00	10,000.00	2,785,000.00	0.00	0.06	928.00	0.00	0.00
Pasteurization	2,000.00	4,000.00	2,785,000.00	19.00	440.96	4,000.00	0.00	11.02
UHT Milk	2,000.00	3,500.00	557,434.00	0.02	0.07	9.00	0.00	0.00
Milk Packaging	2,000.00	2,500.00	557,434.00	0.83	3.85	67.00	0.00	0.15
Sour Milk	2,000.00	1,500.00	278,169.00	13.00	30.13	420.00	0.00	2.01
Yogurt	2,000.00	2,000.00	277,941.00	0.45	1.03	25.00	0.00	0.05
Safa Up	2,000.00	2,000.00	279,226.00	0.00	0.00	4.00	0.00	0.00
Chmnit	2,000.00	4,000.00	277,443.00	0.07	0.16	14.00	0.00	0.00
Choco	2,000.00	2,000.00	419,014.00	0.17	0.59	19.00	0.00	0.03
Sour Milk Up	2,000.00	2,000.00	278,235.00	0.06	0.13	11.00	0.00	0.01
Cheese	2,000.00	2,500.00	417,538.00	0.22	0.76	27.00	0.00	0.03
Packaged products	2,000.00	999,999.00	2,785,000.00	0.00	0.00	1.00	0.00	0.00

Fig. 10. Locations summary

Name	Total Exits	Current Quantity In System	Average Time In System (Min)	Average Time In Move Logic (Min)	Average Time Waiting (Min)	Average Time In Operation (Min)
Milk	63,900.00	100.00	25.85	1.10	0.07	24.68

Fig. 11. Entities summary

and incubator, Choco and inventory and the remaining departments have unimportant relationships (Fig. 13).

After we constructed the relationship chart, we calculated TCR (total closeness rate) [5] for each department. And we discover that incubator has the largest TCR value.

According to the TCR value we could create the placement sequence to construct a new layout design. The placement sequence: incubator-Safeno-maintenance-condensed milk filling-receiving-cleaning-sour milk-Shemaint-roll material-holding tanks 1-electric-condensed Milk-Choco-UHT-sour milk up-inventory-lab-holding tanks 2-fridge-pasterization-raw material-cleaning2. We choose incubator as the first department to construct the new layout design because it has the higher total closeness rating (TCR) which equal 1024. Then we choose Safeno department which has ordinary important relationship with incubator. After that we choose maintenance, condensed, milk filling and receiving and cleaning departments because they have higher number of unimportant relationships with the already placed departments. All of holding tanks, roll materials, cement and sour milk have same number of unimportant relationships with the already placed departments and the same TCR value so we ordered them according to the value of unimportant relationship of each one. For example, the value of unimportant relationship between incubator and Shemaint is 19620. Lab, Sour milk up, inventory, holding tanks 2, condensed milk, Choco, UHT departments don't have relationship with the already placed departments so we choose them according to the value of unimportant relationship they have (Fig. 14).

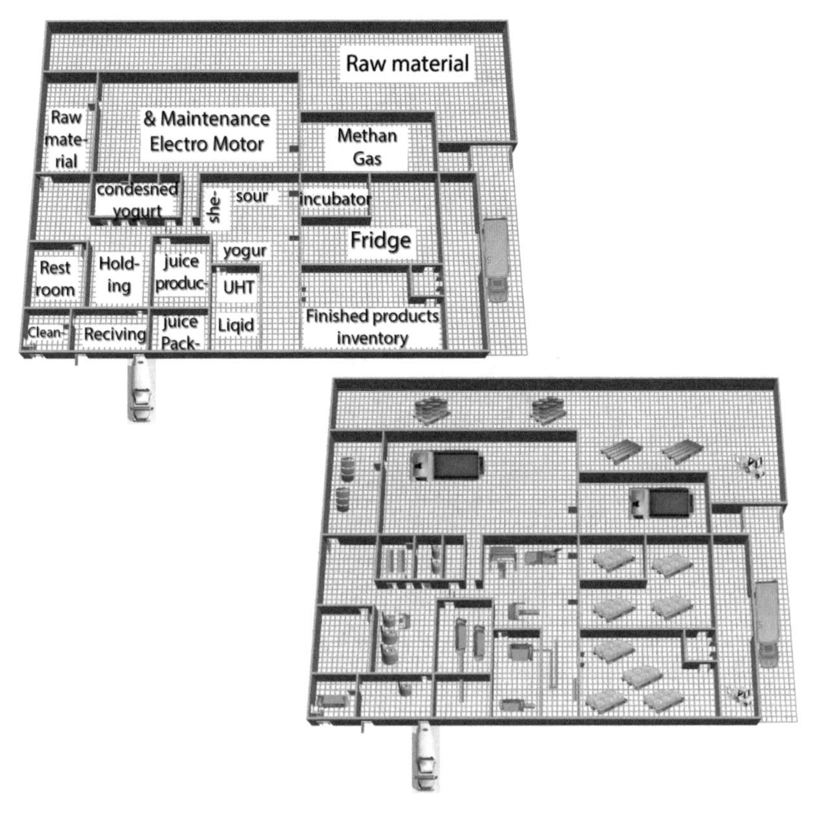

Fig. 12. First floor layout

Second floor:

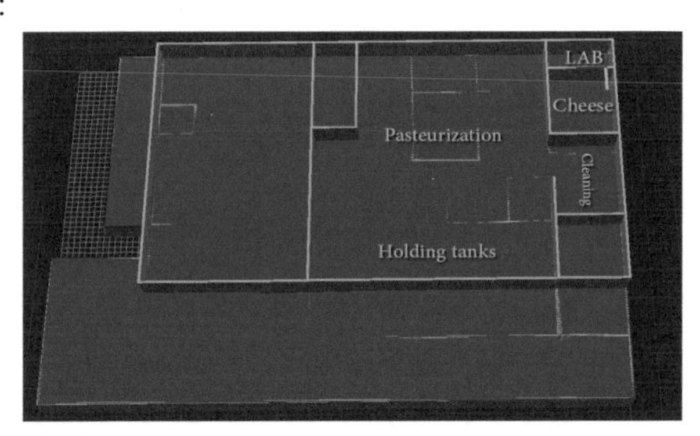

Fig. 13. Second floor layout

This is new layout design that we constructed from **CORELAP** chart depending on highest placing rating which is the sum of weighted closeness ratings between the department to enter the layout and its neighbors (Fig. 15).

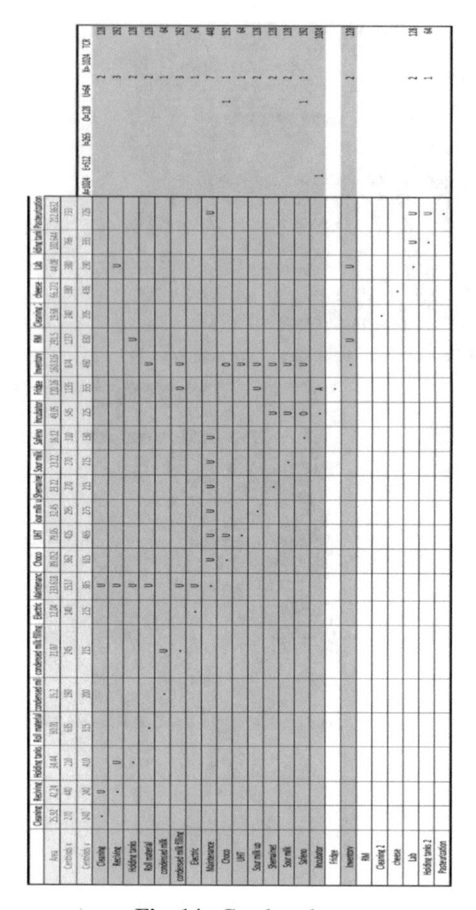

Fig. 14. Corelap chart

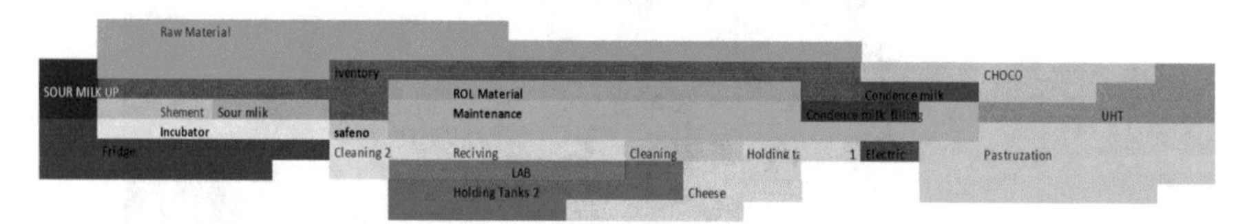

Fig. 15. New corelap layout

We've applied craft technique to improve current layout. At the beginning we evaluate current layout correspond to adjacency-based score by multiplying flow with distance with cost per unit. Cost per trip = operator salary/(number of working hours per day) * (number of working days per week) * (average number of flow) = 2000/(6 * 8 * 32.6) = 1.2 NIS (Table 2).

By implementing the multiple method based on the relationship and the flow between charts we moved the laboratory to the receiving department to have new department "research and development" (Fig. 16)

Table 2. Flow * Distance * Cost table

F*d*c	Departments
19682.4	Cleaning
95402.4	Receiving
722325	Holding tanks
86058.6	Roll material
2916	Condensed milk
570307	Condensed milk filling
17844	Electric
209604	Maintenance
167450.4	Choco
50803	UHT
89342	Sour milk up
229812	Shemaint
157941	Sour Milk
121785	Safeno
707616	Incubator
	Fridge
43312	Inventory
	Raw Material
	Cleaning 2
	Cheese
29524	Lab
5611.2	Holding Tanks 2
	Pasteurization
3307653.6	Total

Evaluating the new adjacency score = 175 is better than the old one of 135.

4 Conclusion

After long search, we found that the factory fulfills our improvement aims due to its pitfalls that need a lot of improvements, so that we can conclude them as follows. Firstly, the flow of materials and labors in the factory is difficult so the layout is no optimize due to that there is a time and cost waste in transportation. Secondly, the factory doesn't have neither the lean manufacturing nor the supply chain principles, so the ability of stopping the production due to the low raw material or packaging supplies. Finally, due to the supervision of human resources department, there is a deficiency in the commitment of

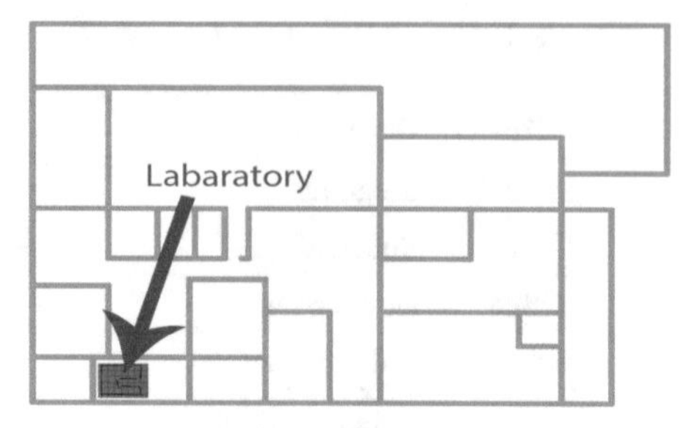

Fig. 16. Multiple method

the workers towards the working hours so that they may came late and left the work early. Moreover, most of the production lines started production lately, due to the long setup time and deficient planning. So, figuring out these problems using flow charts, simulation and plant layout will solve them.

Acknowledgment. This Book Chapter is funded by the International Exchange Program of Harbin Engineering University for Innovation-oriented Talents Cultivation.

References

1. Toffaha, K.M., Dongyan, S.: Diary factory modeling, simulation and layout assessment and improvement. In: Proceedings of The World Congress on Engineering and Computer Science. LNECS, 23–25 October 2018, San Francisco, USA, pp. 752–757 (2018)
2. Tompkins, A.J., White, A.J., Bozer, A.Y., Tanchoco, A.M.J.: Facility Planning, 4th edn. Wiley, New York (2010)
3. Harrell, R.C.: Simulation Using ProModel, 2nd edn. McGraw-Hill, New York (2003)
4. Benjaafar, S., Soewito, A., Sheikhzadeh, M.: Performance Evaluation and Analysis of Distributed Plant Layouts. Working Paper, Department of Mechanical Engineering, University of Minnesota. Minnesota, Minneapolis (1995)
5. Bair, F., Langer, Y., Caprace, J.D., Rigo, P.: Modeling Simulation and Optimization of a Shipbuilding Workshop. Macmillan, New York (2005)

Temperature Distribution and Boundary Condition on Heat Transfer from Discretized Element of Dried Ginger Rhizome Using MATLAB PDES for Optimal Preservation

Austin Ikechukwu Gbasouzor[1](✉), Joshu Depiver[2], and Jude Ebem Njoku[3]

[1] Department of Mechanical Engineering, Chukwuemeka Odumegwu Ojukwu University, P.M.B. 02, Uli, Nigeria
unconditionaldivineventure@yahoo.com, ai.gbasouzor@coou.edu.ng
[2] Department of Mechanical and Manufacturing Engineering and Built Engineering, University of Derby, Britannia Markeaton Street Derby, DE22 3AW, Derby, UK
j.depiver@derby.ac.uk, jdepiver@aol.com
[3] Department of Engineering and Science, University of Greenwich, Medway Campus, Central Avenue Chatham Maritime Kent ME4 4TB, London, UK
jn9744e@gre.ac.uk

Abstract. This paper presents the convective drying of ginger rhizomes under blanched, unblanched, peeled unpeeled condition using matrix laboratory (MATLAB) for finite element the temperature distribution while drying. This research work is an extension of the previous work done with the ARS-680 Environment Chamber for the Drying and TD 10024a-linear heat conduction experimental equipment used in measuring the thermal conductive of the ginger at 6 temperature levels ranging from 10 °C–60 °C and drying time of 2–24 h. The partial differential equation toolbox was employed to PDES for diffusion heat, transfer, structural mechanics, electrostatics, magnetostatics, and ACpower electromagnetics, as well as custom, coupled system of PDES. The discretized meshed of the ginger rhizomes samples have 545 nodes (element) and 1024 (triagles) and the high temperature distribution could be responsible in the colour change obtained for the final product.

Keywords: Blanched · Conductivity · Discretized element · Dried · Ginger rhizomes · Matrix laboratory · Nodes · Peeled unblanched · Unpeeled

1 Introduction

In numerical analysis, finite elements method is a numerical technology for finding approximate solution to partial differential equations and their systems (PDES). The piratical application of finite element method, EFM is known as Finite element analysis, FEA. Finite element analysis has become a common place in recent years that even complicated stress problems can now be obtained with it.

S.-I. Ao et al. (Eds.): WCECS 2018, *Transactions on Engineering Technologies*, pp. 87–100, 2020.
https://doi.org/10.1007/978-981-15-6848-0_8

Convective drying can be employed to remove volatile liquid fropm porous material such as food stuffs, ceramic products, clay products, wood and so on. Porous materials have macroscopic capillaries and pores which cause a mixture of transfer mechanism to occur simultaneously when subjected to heating or cooling. The drying of moist porous solids involves simultaneous heat and mass transfer [1].

Moisture is removed by evaporation into an unsaturated gas phase. Drying is essentially important for preservation of agricultural crops for future use.

Crops are preserved by removing enough moisture from them to avoid decay and spoilage. For example, the principle of the drying process of ginger rhizomes involves decreasing the water content of the product to a lower level so that micro-organisms cannot decompose and multiply in the product. The drying process unfortunately can cause the enzymes present in ginger rhizomes to be killed [2].

Ginger is the rhizome of the plant *Zingiberofficinale*. It is one of the most important and most widely used spices worldwide, consumed whole as a delicacy and medicine. It lends its name to its genus and family *zingiberaceae*. Other notable members of this plant family are turmeric, cardamom, and galangal. Ginger is distributed in tropical and subtropical Asia, Far East Asia and Africa (Fig. 1).

Fig. 1. Fresh ginger rhizomes

Ginger is not known to occur in the truly wild state. It is believed to have originated from Southeast Asia, but was under cultivation from ancient times in India as well as in China. There is no definite information on the primary center of domestication. Because of the easiness with which ginger rhizomes can be transported long distances, it has spread throughout the tropical and subtropical regions in both hemispheres. Ginger is indeed, the most wildly cultivated spice [3]. India with over 30% of the global share, now leads in the global production of ginger.

1.1 Heat Transfer of the Ginger Rhizomes Using MATLAB Partial Differential Equation Toolbox™ (PDE Toolbox) and Computer Programme Developed

The MATLAB Partial Differential Equation Toolbox™ have the capabilities of solving partial differential equations (PDEs) in 2-D, 3-D and time using finite element analysis. It can specify and mesh 2-D and 3-D geometries and formulate boundary conditions and equations. The PDE Toolbox was employed to PDEs for diffusion, heat transfer, structural mechanics, electrostatics, magnetostatics, and AC power electromagnetics, as well as custom, coupled systems of PDEs. In this study, the Boundary condition chosen for the heat transfer problem is the Dirichlet Boundary condition and the PDE specification employed is the elliptic which are mathematically expressed as:

$$\text{Dirichlet Boundary Condition} : hu = r \tag{1}$$

Where h is a matrix, u is the solution vector, and r is a vector.

$$\text{Elliptic PDE specification:} \; - div(k * grad(T) = Q + h * (T_{ext} - T) \tag{2}$$

Where T is temperature, Q is heat source, k is the coefficient of heat condition, h is the convective heat transfer coefficient,T_{ext} is the external temperature.

2 Methodology

2.1 Discretization Method

Partial differential equations (PDEs) are widely used to describe and model physical phenomena in different engineering fields and also in microelectronics fabrications. Finite element formation works on a large number of discretized elements and on different kinds of meshes within the domains. It also provides good results for a coarse mesh.

A computer programme was also developed in MATLAB to easily compute, analyse and conduct simulations for the ginger drying [4, 5].

Figure 2 shows the discretized meshed of the ginger rhizome in line with the cut geometry for the different case under study. The discretized samples have 545 nodes and 1024 triangle elements.

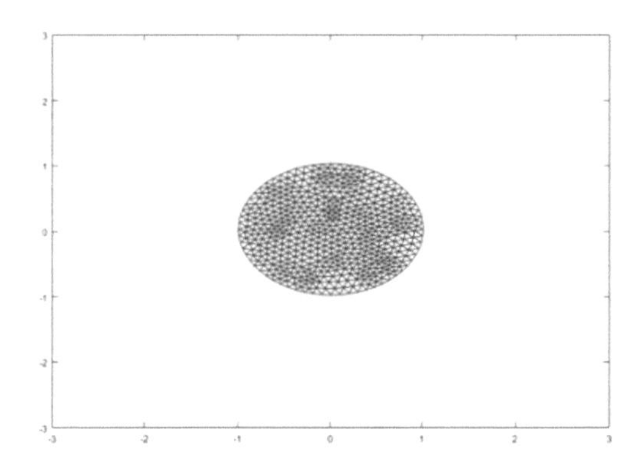

Fig. 2. Discretized mesh with 545 nodes and 1024 triangles

Figures 3 describes the temperature distribution of unblanched, blanched, peeled and unpeeled ginger samples at temperatures: 10 °C, 20 °C, 30 °C, 40 °C, 50 °C and 60 °C. In Fig. 3, the temperature distribution for the unblanched ginger at 10 °C transmits heat radially from 10 °C to a final peak temperature of 60 °C. For the blanched ginger as shown in Fig. 4, the heat is transmitted radially from 10 °C to a final peak temperature of 70 °C. In Fig. 5, at 10 °C for the peeled ginger rhizome. It can be clearly seen that the temperature distribution is radial from 10 °C to a final peak temperature of 60 °C while for the unpeeled ginger rhizomes in Fig. 6, the distribution radiates from 10 °C to a final peak temperature of 70 °C.

Similarly, at a temperature of 20 °C. The temperature distribution in Figs. 7 and 9, the unblanched and peeled rhizomes respectively, looks alike as both figures radiates from 10 °C to a final peak temperature of 60 °C. In contrast, the temperature in Fig. 8 radiates from 10 °C to a final peak temperature of 80 °C while in Fig. 10, the temperature rose steadily from 10 °C to 70 °C.

For the temperature distributions at 30 °C to 60 °C as typified in Figs. 11, 12, 13, 14, 15, 16, 17, 18, 19, 20, 21, 22, 23, 24, 25, and 26, the peak radial temperatures were seen to higher than what was obtained initially at 10 °C and 20 °C. A thorough look in Figs. 11, 12, 13, 14, 15, 16, 17, 18, 19, 20, 21, 22, 23, 24, 25, and 26 show that the temperature distribution at 40 °C was remarkably higher than those obtained at 30 °C and 60 °C but compare relatively to the values obtained at 50 °C. The high temperature distribution could be responsible to the colour change obtained for the final product (Table 1) [6].

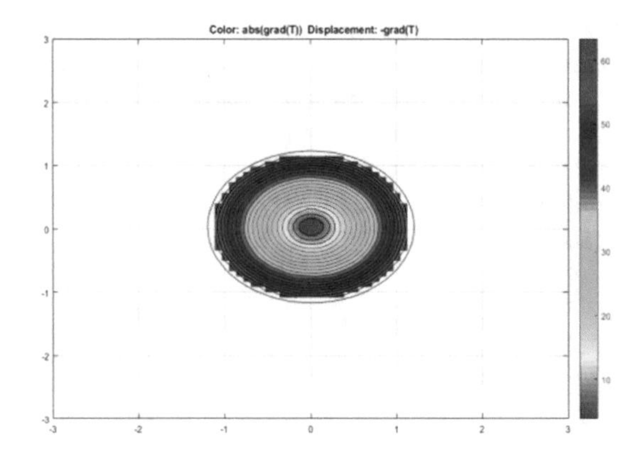

Fig. 3. Temperature distribution for the Unblanched at 10 °C

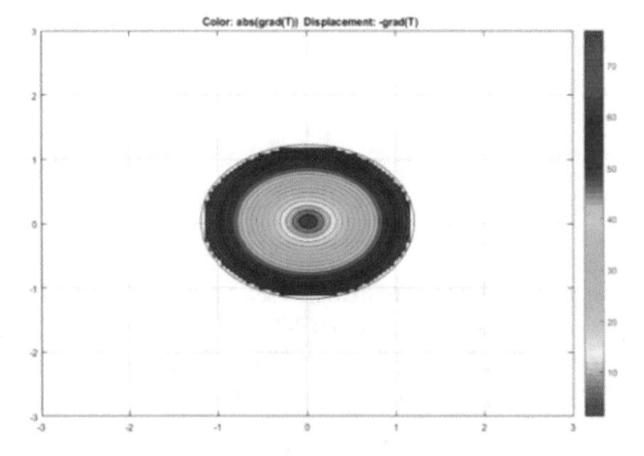

Fig. 4. Temperature distribution for the Blanched at 10 °C

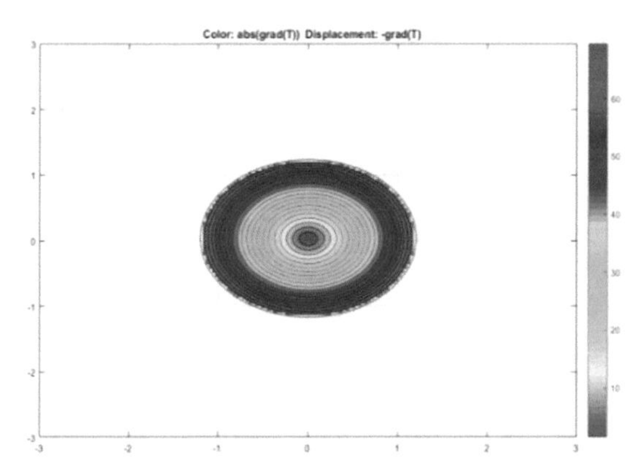

Fig. 5. Temperature distribution for the Peeled at 10 °C

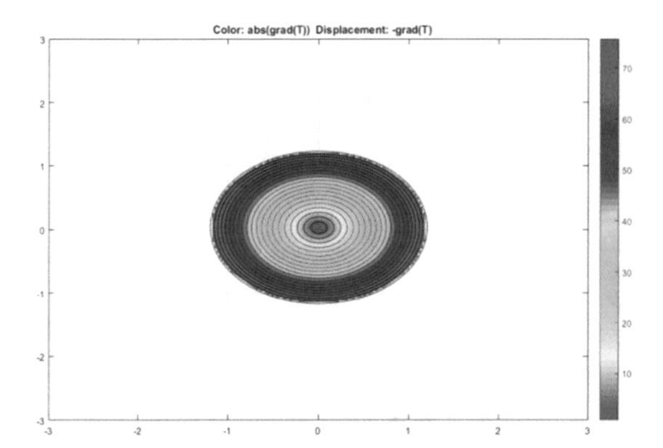

Fig. 6. Temperature distribution for the Unpeeled at 10 °C

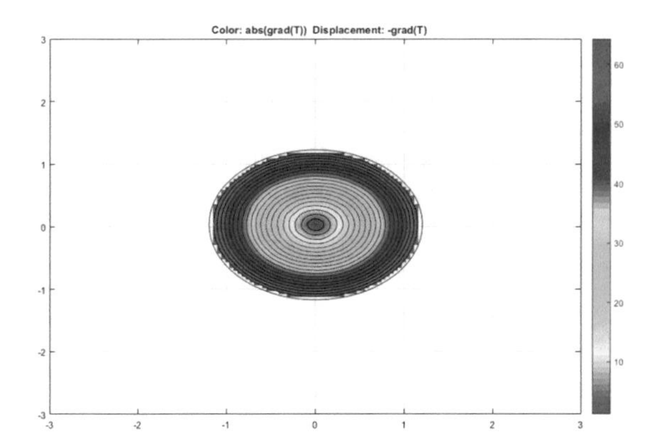

Fig. 7. Temperature distribution for the Unblanched at 20 °C

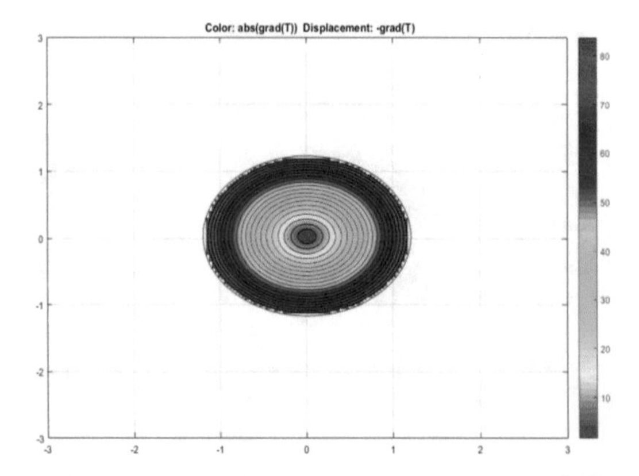

Fig. 8. Temperature distribution for the Blanched at 20 °C

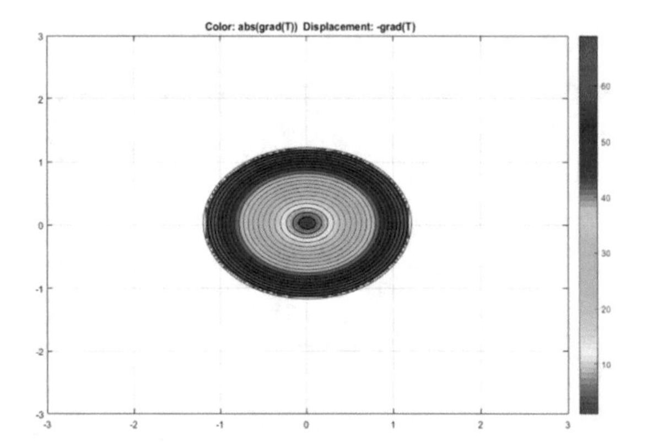

Fig. 9. Temperature distribution for the Peeled at 20 °C

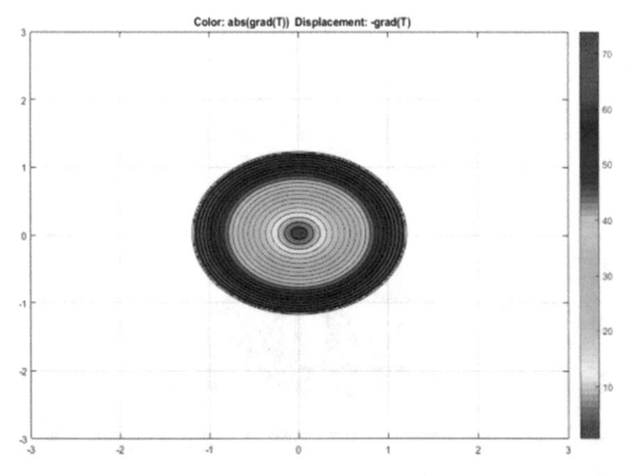

Fig. 10. Temperature distribution for the Unpeeled at 20 °C

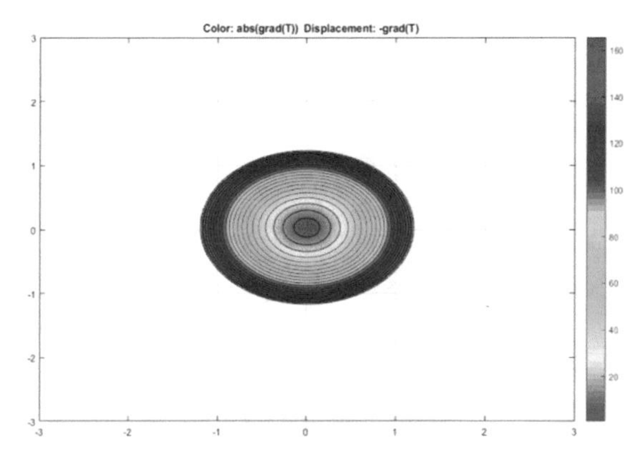

Fig. 11. Temperature distributionfor the unblanched at 30 °C

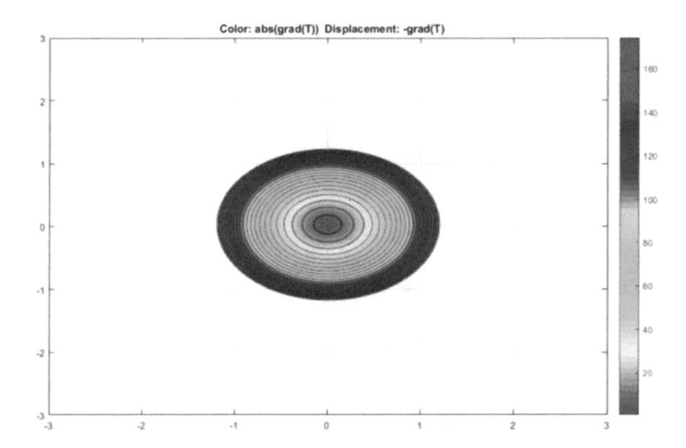

Fig. 12. Temperature distributionfor the Blanched at 30 °C

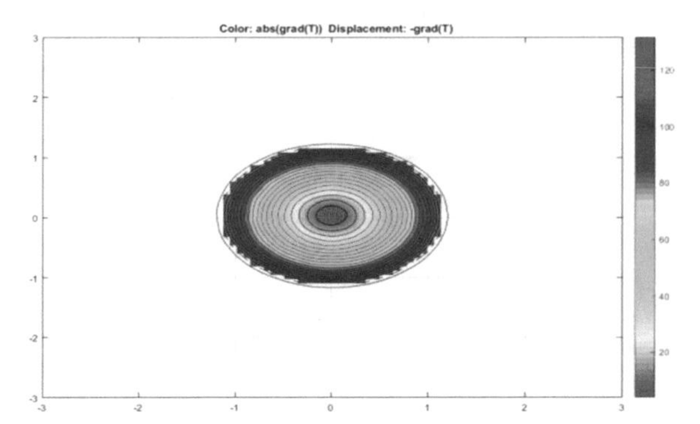

Fig. 13. Temperature distribution for the Peeled at 30 °C

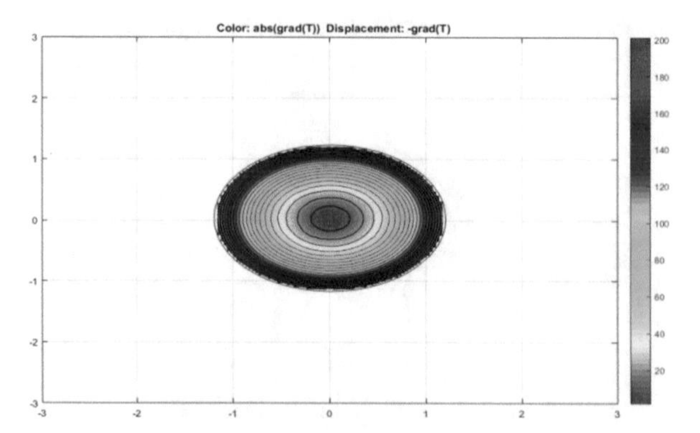

Fig. 14. Temperature distribution for the unpeeled at 30 °C

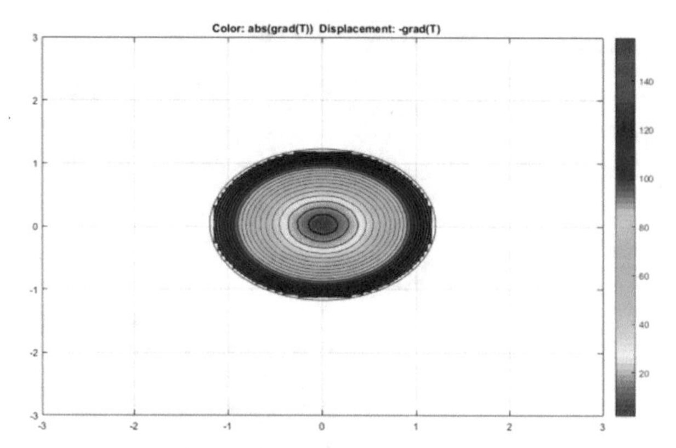

Fig. 15. Temperature distribution for the unblanched at 40 °C

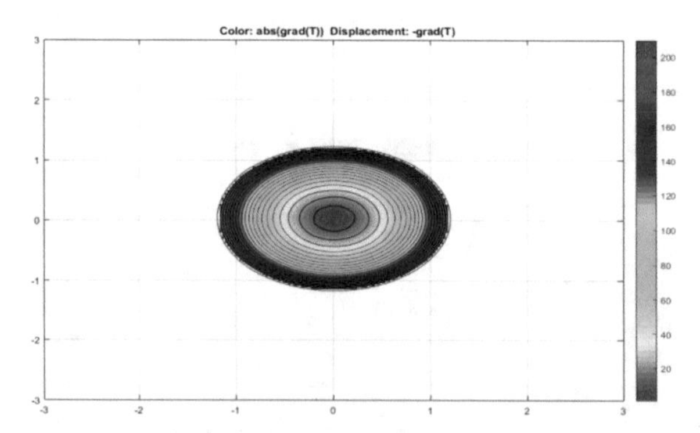

Fig. 16. Temperature distribution for the Blanched at 40 °C

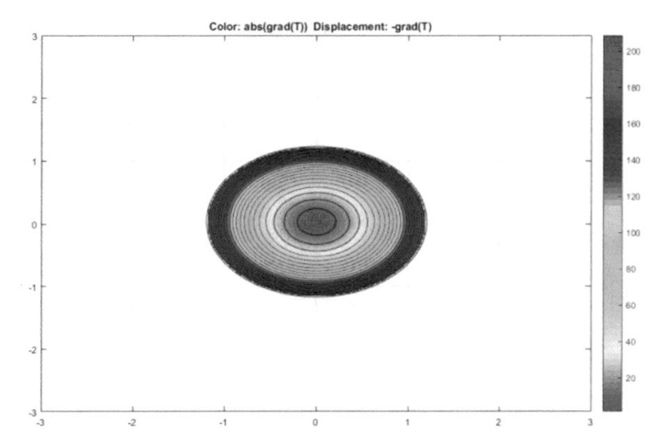

Fig. 17. Temperature distribution for the Peeled at 40 °C

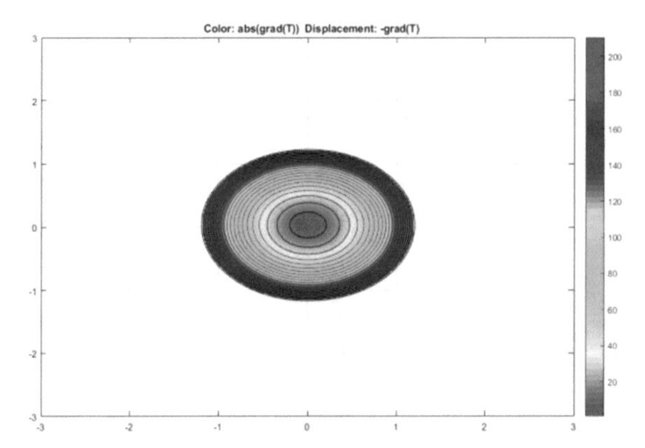

Fig. 18. Temperature distribution for the unpeeled at 40 °C

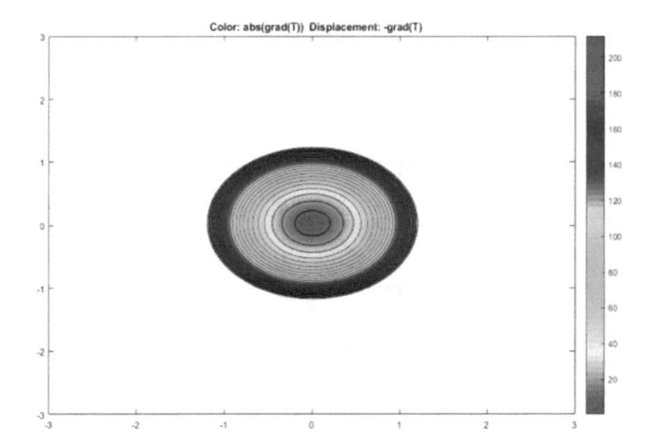

Fig. 19. Temperature distribution for the unblanched at 50 °C

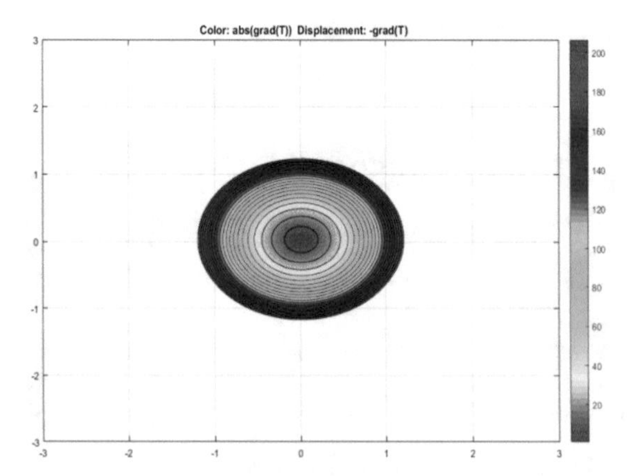

Fig. 20. Temperature distribution for the Blanched at 50 °C

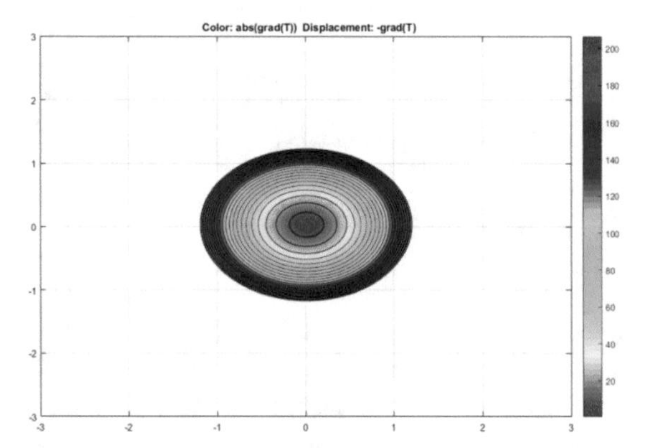

Fig. 21. Temperature distribution for the Peeled at 50 °C

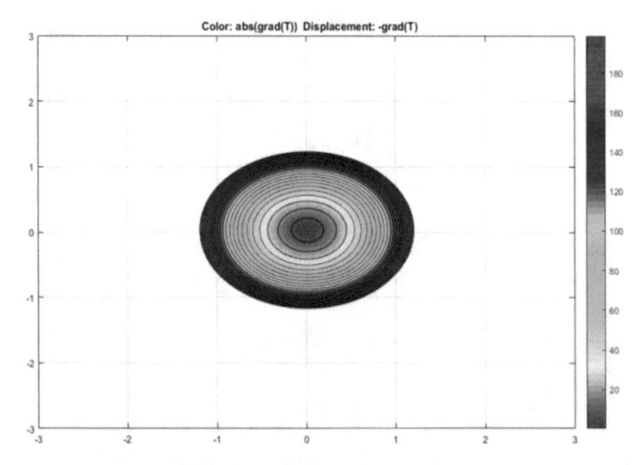

Fig. 22. Temperature distribution for the unpeeled at 50 °C

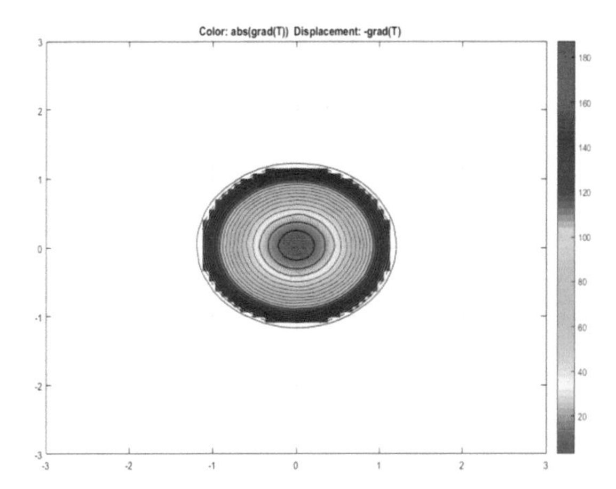

Fig. 23. Temperature distribution for the unblanched at 60 °C

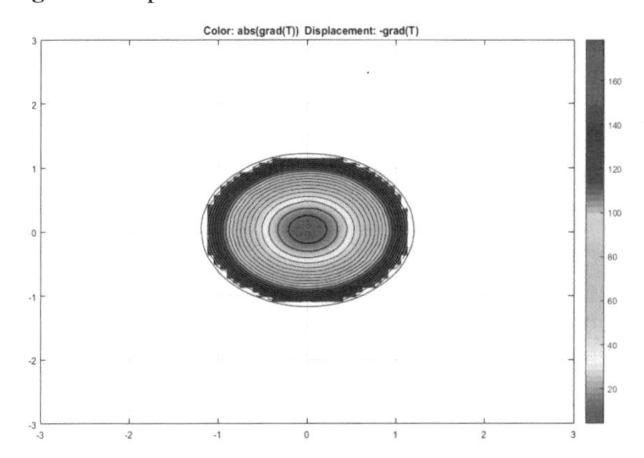

Fig. 24. Temperature distribution for the blanched at 60 °C

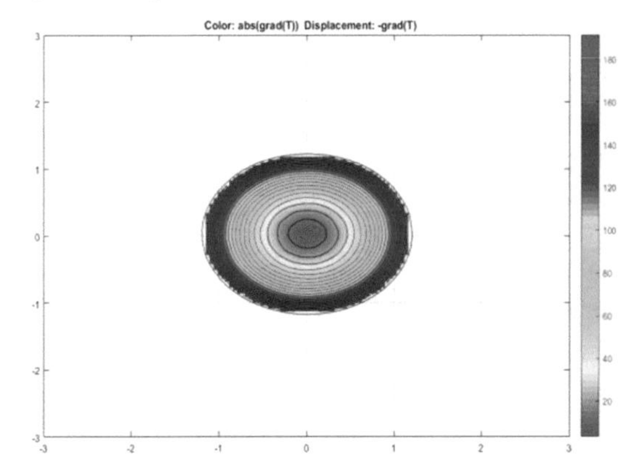

Fig. 25. Temperature distribution for the Peeled at 60 °C

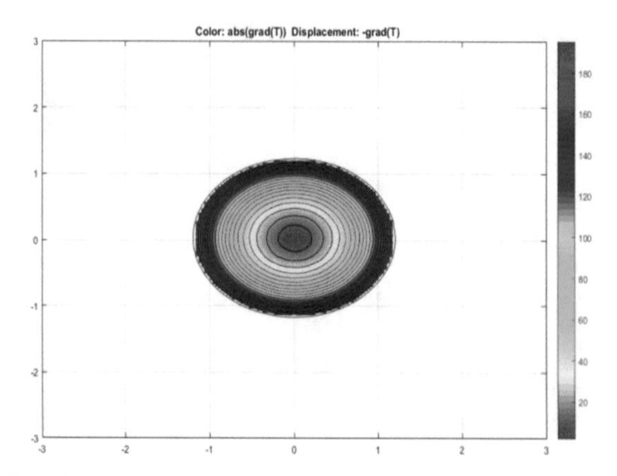

Fig. 26. Temperature distribution for the unpeeled at 60 °C

Table 1. Table of Moisture Content (%) and Thermal Conductivity (W/m.K) for Unblanched, Blanched, Peeled and Unpeeled Ginger Rhizomes from 10 °C to 60 °C to and Drying time of 2 and 24 h [7]

Temperature	10 °C		20 °C		30 °C	
Time (Hour)	2	24	2	24	2	24
Final Moisture Content (%)						
Unblanched	88.84	49.55	86.55	47.81	87.34	39.55
Blanched	84.58	41.13	86.29	34.26	86.65	17.48
Peeled	88.74	55.91	87.85	37.49	87.95	27.76
Unpeeled	91.08	62.22	86.17	48.36	87.71	31.15
Thermal Conductivity (W/m. K)						
Unblanched	0.406	0.161	0.406	0.149	0.107	0.068
Blanched	0.329	0.140	0.292	0.131	0.1006	0.069
Peeled	0.377	0.143	0.377	0.139	0.1459	0.065
Unpeeled	0.340	0.171	0.345	0.171	0.1126	0.061
Temperature	40 °C		50 °C		60 °C	
Time (Hour)	2	24	2	24	2	24
Final Moisture Content (%)						
Unblanched	79.32	30.12	71.65	17.95	74.16	6.63
Blanched	70.11	17.00	66.64	10.25	63.11	9.04
Peeled	75.93	23.92	65.50	13.21	70.75	8.56
Unpeeled	81.46	26.30	67.85	15.49	74.36	5.98
Thermal Conductivity (W/m. K)						
Unblanched	0.076	0.056	0.072	0.054	0.076	0.055
Blanched	0.071	0.056	0.073	0.0556	0.084	0.052
Peeled	0.072	0.052	0.076	0.0519	0.079	0.048
Unpeeled	0.072	0.054	0.078	0.0460	0.078	0.046

3 Conclusion

In this study the following conclusion was drawn from this study:

- In the study of moisture content of ginger rhizomes, it was deduced that ginger could be dried at different temperatures. Ginger rhizomes dried at short time/low temperature will not reduce the effects of past and bacterial infections, but when dried at high temperature say 60 °c will drastically reduced the effects of pest and bacterial infecting associated with moist ginger rhizomes
- The drying rate at higher drying times (24 h) was 0.889/°C and 0.4437/°C for 2 h drying, giving 50% in moisture reduction rate. The intercept which theoretically gives the initial moisture content at 0 °C is lower at 24 h drying (59.33%) compared to 95.12% on dry basis at 2 h of drying, as expected. In Afolayan M.O et al. [8].
- The result of this study shows that the lowest moisture content (5.98%) is obtained for unpeeled ginger while the highest is the blanched (9.04%) all for 24 h-drying and at 60 °C.
- The average moisture content for 2 h drying at 60 °C was 70.6% while for 24 h drying, it was an average of 7.55%. which is close to the range of 4–7% as expected in Eze, J and Agbo, K (2011) This is better than the result of 22.54% obtained at 50 °C under blanched condition drying for 32 h (Hoque *et al.*, 2013).
- At higher temperatures ginger shrinkages and surface decolouration may occur. good results are achievable at temperature of 60 °C to sustain the quality of the products for preservation.
- The thermal conductivity for 24 h-dried ginger at 60 °C approximates to the thermal conductivity of dried ginger and it is 0.050 W/mK on the average, with unpeeled ginger giving the lowest value of 0.046 W/mK and unblanched ginger giving the highest value of 0.055 W/mK.
- MATLAB gave good approximations by Finite Element method the temperature distributions within the ginger rhizomes at different drying temperatures.
- The essence of testing for thermal conductivities of samples is to gain proper understanding of their thermal behaviours. They are also utilized for the modelling and development of processes such as drying and freezing as mass and energy are exchanged. Other thermal properties such as thermal diffusivity and specific heat were not included in this research.
- The drying curve for the Nigerian ginger rhizomes under all conditions of temperature and time shows that increasing rate of drying behaviours clearly shows that maximum drying takes place at a higher temperature and time such as when external factors such as moisture barriers and air/mass movement influencing the drying process. The drying process is relatively slow at lower temperatures of 10 °C–20 °C (Figs. 3, 4, 5, 6, 7, 8, 9, 10, 11, 12, 13, 14, 15, 16, 17, 18, 19, 20, 21, 22, 23, 24, 25, and 26) and increases from 30 °C–60 °C with an increase of drying time from 14–24 h.
- Temperature distribution of discretized elements applied on the dried ginger rhizomes using MATLAB PDES for optimal preservation of ginger rhizomes.

Acknowledgement. The authors acknowledges the Vice Chancellor Prof. Greg Chukwudi Nwakoby of Chukwuemeka Odumegwu Ojukwu University, Anambra State, Nigeria and TETFUND for support and financial help received throughout the research work in form of grant from the TETFUND. Also to Asso. Prof. Sabuj Mallik a Lecturer in the Department of Mechanical Engineering and Built Environment College of Engineering & Technology University of Derby/Britannia, for his support, and assistance particularly my co-author in persons of Jude & Joshua.

References

1. Hoque, M., Bala, B., Hossain, M., Uddin, M.B.: Drying kinetics of ginger rhizome (Zingiber officinale). Bangladesh J. Agric. Res. **38**(2), 301–319 (2013). https://doi.org/10.3329/bjar.v38i2.15892
2. Omeni, B.: About Us: Agronigeria Ltd. (Agronigeria, Producer, & Agronigeria Ltd) (2015). An Agronigeria Ltd Website: http://agronigeria.com.ng/2014/01/09/steps-to-take-in-ginger-plantation. Accessed 7 Mar 2015
3. Lawrence, B.M.: Major Tropical Spices-Ginger (Zingiber officinale Rose). Perfumer Flavorist **9**, 1–40 (1984)
4. Ikechukwu, G.A., Depiver, J., Njoku, J.E.: Application of temperature distribution on discretized element of dried ginger using MATLAB PDES for optimal preservation. In: Proceedings of the World Congress on Engineering and Computer Science. LNECS, 23–25 October 2018, San Francisco, USA, pp. 712–718 (2018)
5. Gbasouzor, A.I.: Prediction of thin layer drying characteristics of ginger rhizome slices in convective environment. Ph.D. thesis, Department of Mechanical Engineering, Nnamdi Azikiwe University, Awka, Nigeria (2019)
6. Gbasouzor, A.I., Omenyi, S.N., Mallik, S.: Simulation and discretized element of ginger on thin layer Drying using MATLAB PDEs for optimal programming. J. Eng. Appl. Sci. **16**(1) (2020)
7. Ikechukwu, G.A., Omenyi, S.N.: Convective drying of ginger rhizomes. In: Ao, S.I., Kim, H., Amouzegar M. (eds) Transactions on Engineering Technologies. WCECS 2017, pp. 83–98. Springer, Singapore (2019). https://doi.org/10.1007/978-981-13-2191-7_7
8. Afolayan, M.O., Adama, K., Oberafo, A., Omojola, M., Thomas, S.: Isolation and characterization studies of ginger (Zingiber officinale) root starch as a potential industrial biomaterial. Am. J. Mat. Sci. **4**(2), 97–102 (2014). Accessed 12 March 2015. http://article.sapub.org/10.5923.j.materials.20140402.06.html

Molecular Docking Analysis: Interaction Studies of Natural Compounds with Human TG2 Protein

Prachi P. Parvatikar(✉) and Shivkumar B. Madagi

Department of P.G. Studies and Research in Bioinformatics, Karnataka State Akkamahadevi Women's University, Vijaypur, Karnataka, India
prachisandeepk@gmail.com, madagisb@gmail.com

Abstract. Transglutaminase 2 (TG2) is a multi-domain, multi-functional and ubiquitously expressed proteins. It plays a role in diverse biological functions, while abnormal level of this protein is believed to be involved in the pathogenesis of several diseases including lung cancer. TG2 act as a good therapeutic target against lung cancer. A variety of bioactive compounds from medicinal plants have been reported to possess anti-cancer as well as anti-tumor properties. In present study natural compound from different classes were subjected for screening ADME/T properties and molecular docking analysis to investigate their interaction modes with the potential TG2 target. The result revealed that Artobiloxanthone have the highest affinity to TG2 and the highest consistency of interaction model.

Keywords: ADME/T · Aliphatic · Anti-cancer · Cancer · Flavonoids · Molecular docking · Simple aromatic · TG2

1 Introduction

Transglutaminases is nine members family which were first discovered in the 1950s isolated from mammalian liver homogenates. This family includes transglutaminases 1–7, factor XIIIA, and the enzymatically inactive erythrocyte band 4.2 [1]. They all are facilitates calcium-dependent covalent bonds formation between small molecule amines and releases ammonia [2]. Among this transglutaminase 2 (tissue transglutaminase, TG2) is extensively studied and biologically characterized protein. It is expressed in most of body tissue, plays various biological roles [3]. TG2 has been connected to a number of diseases, such as celiac sprue, Alzheimer's disease, Huntington's disease [4], and certain types of cancer [5].

Transglutaminase 2 is an ~80 kDa protein multidomain G-protein. It catalyses the calcium-dependent covalent modification of protein by either transamidation or deamidation reactions. Histological staining of TG2 from mammalian organs has revealed that this protein is ubiquitous expressed throughout the body [6]. Among the few common cell types are endothelial cells, macrophages, fibroblasts, smooth muscle cells, a fraction of extracellular TG2 found in the ECM [7].

S.-I. Ao et al. (Eds.): WCECS 2018, *Transactions on Engineering Technologies*, pp. 101–111, 2020.
https://doi.org/10.1007/978-981-15-6848-0_9

Many biological activities have been played by TG2 such as wound healing [8], macrophage phagocytosis [9], TGF-β activation, NF-κB activation [9], protein kinase activity [10], and association with G-protein coupled receptor GPR56 [11].

The biological function of TG2 is dependent up conformational changes in protein. Recently, two distinct conformations in human TG2 have been characterized via x-ray crystallography, one with GDP bound and the other with an active site covalent inhibitor bound to it [12]. Transglutaminase 2 consists of four distinct domains: 1) an N-terminal β-sandwich domain that contains the fibronectin binding site, 2) the catalytic core domain composed of interspersed α-helices and β-sheets containing the substrate binding pocket and catalytic triad 3) a β-barrel domain with a binding pocket for GTP and 4) a C-terminal β-barrel [13].

1.1 Transglutaminase 2 in Cancer

A number study has revealed that TG2 plays an important role in the development of certain types of cancer. It was also shown that TG2 protein is upregulated in cancerous tissue relative to healthy tissue in cancers such as glioblastomas [14], malignant melanomas, and pancreatic ductal adenocarcinomas [15]. Apart from that there is positive correlation between the chemotherapeutic resistance and metastatic potential of certain cancers with TG2 expression levels [16]. TG2 also play as anti-apoptotic protein in certain cell type. However certain study also suggest that the down-regulation of TG2 expression contribute to development of some cancer. The significance of TG2 to cancer biology may depend upon the cancer cell type, the type of cancer, the location of the cancer, and possibly the stage of the cancer. The study suggest that TG2 strongly promotes the oncogenesis by promoting the activation NF-κB by intracellular cross linking which in turn activates expression of anti-apoptotic proteins such as Bcl-xL and BFL [17].

Currently small molecule and peptidomimetic inhibitors which are capable of blocking TG2 enzymatic activity have been designed and biochemically characterized to treat patients that have these devastating and often fatal diseases because of unusual expression of TG2 [18]. However, their benefits are limited by the variety of systemic side effects and the development of resistance after chronic use. Thus, developing new drug candidates from natural products is greatly interesting.

1.2 Anti-cancer Activity of Natural Compounds

Medicinal plants contain many natural compounds, such as flavonoids, terpenoids, alkaloids, and saponin. They have been reported to have in vitro as well as in vivo anti-cancer activity.

Flavonoid and Phenolic Compounds

Flavonoids are polyphenolic compounds that are ubiquitously in plants. They have been shown to possess a variety of biological activities at nontoxic concentrations in organisms. Flavonoids present as different derivatives such as aglycone, glycosides. They are grouped as flavonols, flavones, cathecins, flavanones, anthocyanidins, and isoflavonoids [19]. The study revealed that it exhibits Varity of anticancer effects such as cell growth and kinase activity inhibition, apoptosis induction, and of tumor invasive behaviour.

Several dietatry flavonoids inhibit the growth of tumor cells [20], and induce cell differentiation. Earlier reports indicate that certain dietary flavonoids also exhibit potent antitumor activity in vivo.

Terpenoids

Terpenoids important constituent of plant. They provide defense against environmental stress and repair wound and injuries. Terpenoids categorized into hemi-, mono-, sesqui-, di-, sester-, tri- and tetraterpenoids. A large numbers of terpenoids have been tested for anti-cancer properties. The anti-cancer activity of terpenoids appears promising and will potentially open more opportunities for cancer therapy [21].

Alkaloid

Alkaloids are the nitrogen containing cyclic compound. They proved to be potent anti cancer substance. Isoquinoline, quinoline, and Indole alkaloids were the most studied classes. Berberine, an Isoquinoline alkaloid, showed potential in vitro and in vivo antitumor activity [22].

1.3 Molecular Docking

Molecular docking is an important tool in drug discovery as well as structural molecular biology. The purpose of docking is to predict binding pattern of ligand with 3-D structure of protein. It finds out the scoring function of correctly docked complex also high dimension space [23]. Molecular docking analysis is useful to understand the interaction of natural compounds with molecular target of anti-tumour activity. Further, the QSAR can be used to develop new derivative natural compounds with higher anti-inflammatory activity. The mechanism of action and molecular target of various natural compounds needs to be studied for constructing a structure activity relationship. The present study is focused on to determine the model of interactions between the selected natural compounds with TG2 as target molecule by molecular docking analysis in Autodock 4.2.

2 Methodology

2.1 Molecular Docking Analysis

Molecular modeling investigations were carried out using Dell, Intel Core i3-4030U Processor, 2 GB RAM, 500 GB hard disk, and Intel HD Graphic Family graphics card. Autodock 4.2 docking program, Molecular Graphic Lab, The Scripps Research Institute was employed for the docking studies.

2.2 Preparation of Target Proteins

PDB structure of human TG2 protein used, i.e., 4PYG was obtained from the Brookhaven Protein Data Bank. The protein structure was prepared using Accerly Discovery studio to remove all non receptor atom including water, ion, and miscellaneous compounds. The obtained structure then was saved as pdb file (Fig. 1).

Fig. 1. Crystallographic structure of human TG2 protein.

2.3 Binding Site Prediction

Active site analysis calculates the number, boundary of mouth openings of every pocket, molecular reachable surface and area. Binding site analysis provides a significant insight to recognize the surface structural pockets, active site, shape and volume of every pocket, internal cavities of proteins. The probable binding site was searched based on structural association of template, PDB sum and literature (Fig. 2).

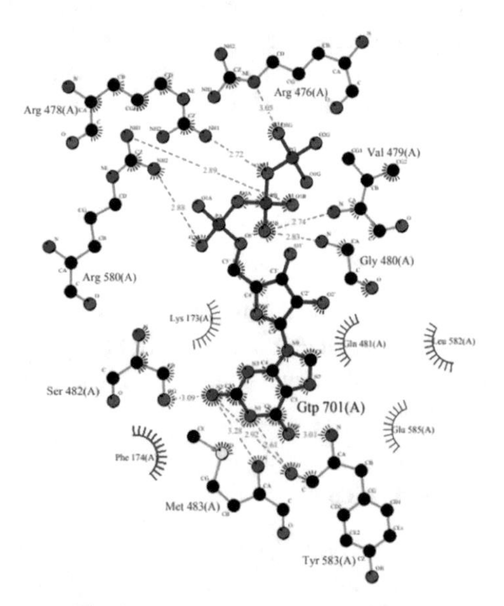

Fig. 2. Binding site residues of TG2.

2.4 Preparation of Ligands

The sdf structure of selected 30 natural compound were retrived from NPACT database [24]. The structures of the ligands were sketched using Marvin sketch. Each structure then was executed for energy minimization. These obtained conformations were saved in pdb file and further used as starting conformations to perform docking analysis (Fig. 3).

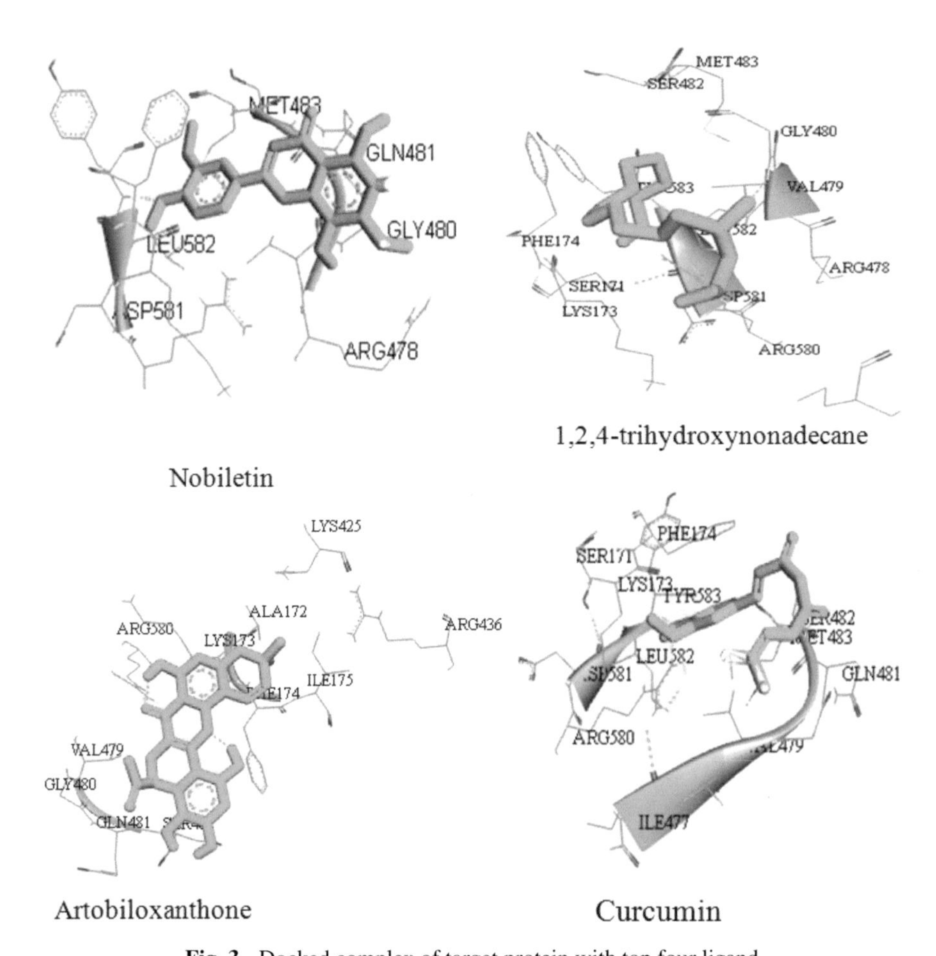

Fig. 3. Docked complex of target protein with top four ligand.

2.5 ADME/T Studies of Compounds

ADME study is essential is primary step of drug compound. It include properties of structural analogues, it predicts both physically significant descriptors and pharmaceutically relevant properties. The properties were predicted using SWISS ADME tool. It consists of principle descriptors and physiochemical properties with a detailed analysis of the log P (Octanol/Water), log S, molecular weight etc. It also calculate the analogues depending upon on Lipinski's rule of 5, which is important step for rational drug design (Table 1).

Table 1. ADMET property of selected ligand.

Drug name	Category	Molecular weight (g/mol)	H Bond donar	H Bond acceptor	Lipinski rule	Log P
Nobiletin	Flavonoids	402.39	0	8	Yes	3.96
Curcumin	Flavonoids	368.38	2	6	Yes	3.27
1,2,4-trihydroxynonadecane	Aliphatic natural	346.57	3	3	Yes	4.50
Artobiloxanthone	Flavonoids	434.444	4	7		3.49
4-Gingerol	Simple aromatic	266.33	2	4	Yes	2.38
3′-formyl-2′, 4′, 6′-trihydroxy-5′-methyldihydrochalcone	Flavonoids	300.31	3	5	Yes	1.76
2′, 4′-Dihydroxy-6′-methoxy-3′, 5′-dimethylchalcone	Flavonoids	298.338	2	4	Yes	2.64
4′-bromoflavone	Flavonoids	301.139	2	1	Yes	2.49
Acacetin	Flavonoids	284.267	5	2	Yes	2.56
Apigenin	Flavonoids	432.381	10	2	Yes	2.17
Garcimangosxanthone A	Flavonoids	342.34	2	6	Yes	3.19
12-Hydroxychiloscyphone	Flavonoids	234.339	2	3	Yes	2.53
Butein	Flavonoids	272.256	4	5	Yes	1.66
Delphinidin	Flavonoids	338.696	6	7	Yes	−5.36
Emoroidocarpan	Flavonoids	350.37	0	5	Yes	3.69
Isobonducellin	Flavonoids	282.295	1	4	Yes	2.59
Kaempferol	Flavonoids	286.239	6	4	Yes	1.74
Limonin	Flavonoids	470.518	0	8	Yes	2.87
Malvidin	Flavonoids	331.3	4	7	Yes	−1.80
Ponciretin	Flavonoids	285.275	1	5	Yes	1.67
Retusin	Flavonoids	358.346	1	7	Yes	3.58
Eucalyptol (T)	Terpenoids	154.253	0	1	Yes	2.53
Menthol	Terpenoids	156.269	1	1	Yes	2.55
Rosmanol	Terpenoids	346.423	3	5	Yes	2.50
Tagitinin C	Terpenoids	348.395	1	6	Yes	2.33
Limonene	Terpenoids	136.238	0	0	Yes	2.72
Vulgarin	Terpenoids	264.32	1	4	Yes	1.86
Weisiensin A	Terpenoids	492.56	2	9	Yes	3.01
Papaverine	Alkaloids	375.85	0	5		0
Tomatidenol	Alkaloids	413.64	2	3	Yes	4.25

2.6 Docking Studies

Docking studies were carried out using prepared target macromolecule and natural compound ligand by employing Autodock 4.2 program [25]. Docking was performed to obtain a population of possible conformations and orientations for the ligand at the binding site. A PDBQT file is created that contains a protein structure with hydrogen in all polar residues. All bonds of ligands were set as rotatable. All calculations for protein-fixed ligand-flexible docking were done using the Lamarckian Genetic algorithm (LGA) method. The docking site on protein target was defined by establishing a grid box with default grid spacing, centred on the position of native ligand. The best conformation was chosen with the lowest binding energy, after the docking search completed. The interactions complex protein-ligand conformations, including hydrogen bonds and the bond lengths were analyzed using Discovery Studio Visualizer 16.1.0.15350.

3 Result and Discussion

3.1 Structure of Target Protein

The structure of TG2 (PDB ID: 4PYG) which was retrieved from PDB.It was determined by X-ray crystallography with a bond length of 0.019 Å and bond angles of 1.966°. The structure also contains 29, 223 number of reflections used and protein of 16, 257 respectively.

3.2 Binding Site Determination

The binding site information of TG2 structure was predicted by performing PDBsum and literature evidences PubMed. The binding site region of TG2 contains 13 amino acid residues. Lys 173, Phe 174, Arg 478, Arg 476, Val 479, Gly 480, Gln 481, Ser 482, Met 483, Arg 580, Leu 582, Tyr 583, Glu 585 were selected from PDBsum as binding site cavity in the current study.

3.3 ADME/T Studies of Ligand

The compounds which are labeled as druglike resemble the existing drug molecules and exhibit the following property cut-off values. The compounds exceeding the cut-off values tend to have solubility and permeability problems which would lead to poor oral bioavailabity. In addition to the rule of five there are other properties which help in distinguishing drug and non drug-like compounds (Table 1). The all 30 ligand selected for the present study showed minimal range of values without toxic groups and obeying all pharmacological properties.

3.4 Molecular Docking Analysis

The molecular docking analysis was performed using Autodock 4.2 program with 40 plant derived compounds, including flavonoids alkaloid, terpenoids. Molecular docking helps in understanding the interaction between target and drug. It determines the lowest energy pose. The molecule is consider more stable when it is having least binding energy. From each of the docked compounds, one bound pose was selected based on good hydrogen bond interactions, binding energy score and ligand energy.

Among 30 molecules Artobiloxanthone, Nobiletin, 1,2,4-trihydroxynonadecane showed good inhibitory activity and formed stable docked complex.

Table 2. Docking results of compounds with TG2.

Drug name	Interacting amino acid	Binding engery	Ligand effciency	Inhibition constant
Nobiletin	MET583, LEU582, ARG482	−6.14	−0.21	54.01
Curcumin	SER482, ARG580, TYR583	−7.31	−0.27	4.41
1,2,4-trihydroxynonadecane	ARG 476, GLU 579, ARG 580.	−5.86	−0.29	50.71
Artobiloxanthone	SER482, ARG580	−5.37	−0.17	115.75
4-Gingerol	ASP 405, GLU579, SER419, ARG 317, ASP581	−5.82	−0.31	31.62
3′-formyl-2′, 4′, 6′-trihydroxy-5′-methyldihydrochalcone	LYS444, ILE416, SER 415, VAL422	−5.86	−0.24	4.15
2′, 4′-Dihydroxy-6′-methoxy-3′, 5′-dimethylchalcone	MET483, TYR583	−6.56	−0.36	15.65
4′-bromoflavone	ARG580	−6.02	−0.33	38.96
Acacetin	MET483, ARG580, TYR583	−6.82	−.032	10.52
Apigenin	MET 483ARG580, TYR583	−6.64	−0.33	13.67
Garcimangosxanthone A	MET483, ARG580	−6.46	−0.26	18.33
12-Hydroxychiloscyphone	SER482	−6.27	−0.37	25.5
Butein	MET483, ARG580, TYR583	−6.42	−0.32	19.74
Delphinidin	TYR583	−6.17	−0.28	30.06
Emoroidocarpan		−6.17	−0.26	11.53
Isobonducellin	SER482, ARG580	−6.35	−0.3	22.25

(*continued*)

Table 2. (*continued*)

Drug name	Interacting amino acid	Binding engery	Ligand effciency	Inhibition constant
Kaempferol	MET482, ARG580, TYR583	−6.51	−0.31	16.83
Limonin	LYS173, ARG476, ARG478, SER482	−7.04	−0.21	6.96
Malvidin	MET482, TYR583	−7.43	−0.31	3.61
Ponciretin	MET483, ARG580, TYR583	−6.79	−0.32	10.63
Retusin	MET483, ARG580	−5.85	−0.22	51.72
Eucalyptol (T)	ARG580	−5.61	0.51	77.36
Menthol	TYR583	−6.18	−0.56	29.69
Rosmanol	LYS425	−6.32	−0.25	23.4
Tagitinin C	SER482, ARG580,	−6.66	−0.27	13.22
Limonene	MET480	−5.43	−0.54	105.38
Papaverine	ARG580, TYR583	−6.44	−0.26	18.89
Tomatidenol	LYS425, ARG436	−6.51	−0.22	17.03
Vulgarin	ARG580	−6.98	−0.37	7.67
WeisiensinA	ARG476, ARG580	−2.83	−0.08	8.36

The human TG2-Artobiloxanthone complex showed binding energy score of –6.14 and inhibition constant 115.75. The amino acid residue SER482 and ARG580f ormed a hydrogen bond and a π π-cation with human TG2. The residues Lys 173, Phe 174, Arg 478, Arg 476, Val 479, Gly 480, Gln 481, Ser 482, Met 483, Arg 580, Leu 582, Tyr 583, Glu 585 involved in molecular interactions of docking studies were correlated with the binding site residues of human TG2.

The human TG2-Nobiletin complex showed binding score of –6.14 kcal/mol with inhibition constant 54.01. The amino acid residues MET583, LEU582 and ARG482 formed three hydrogen bonds with human TG2.

The human TG2-1,2,4-trihydroxynonadecane complex showed binding score of –5.86 kcal/mol and inhibition constant 50.71 kcal/mol. The amino acid residues ARG 476, GLU 579, and ARG 580 formed three hydrogen bonds with human TG2. The residue Phe 174 formed π π-cation with human TG2.

However, human TG2-Malvidin complex performed very poor as compare all other ligand. This complex has low inhibition constant 3.61 and binding engery –7.43. It forms two hydrogen bond MET482 and TYR583 (Table 2).

4 Conclusion

In conclusion, molecular docking studies suggested that the most active compounds Artobiloxanthone and Nobiletin were positioned within the active sites of TG2 with highest binding affinity as compare to others. Those can be used as potential drug targets especially for inhibition of TG2 and treatment of lung cancer. The combined approach of ADME/T and docking used in this work helps in expressing the binding affinity of drug target in the receptor well and also validates as potential candidates for second generation drug target discovery. Development of effective and selective inhibitors of Transglutaminase 2 will make possible elucidation of TG2's role in a lung cancer, which may ultimately lead to effective clinical treatments for same. Thus natural compounds may be alternative to synthetic drug as it has no side effect. In the current work, we outlined the identification of a selective TG2 inhibitor from medicinal plant that will be subjected to further computationally assisted structural optimization to improve its potency, selectivity and validation by *in-vitro* assay.

Conflict of Interest. The author has no conflict of interest to declare.

References

1. Lorand, L., Graham, R.M.: Transglutaminases: crosslinking enzymes with pleiotropic functions. Nat. Rev. Mol. Cell Biol. **4**(2), 140 (2003)
2. Borsook, H., Deasy, C.L., Haagen-Smit, A.J., Keighley, G., Lowy, P.H.: The incorporation of labeled lysine into the proteins of guinea pig liver homogenate. J. Biol. Chem. **179**(2), 689–704 (1949)
3. Fesus, L., Thomazy, V., Autuori, F., Ceru, M.P., Tarcsa, E., Piacentini, M.: Apoptotic hepatocytes become insoluble in detergents and chaotropic agents as a result of transglutaminase action. FEBS Lett. **245**(1–2), 150–154 (1989)
4. Arentz-Hansen, H., Körner, R., Molberg, Ø., Quarsten, H., Vader, Willemijn, MC Kooy, Y., EA Lundin, K., et al.: The intestinal T cell response to α-gliadin in adult celiac disease is focused on a single deamidated glutamine targeted by tissue transglutaminase. J. Exp. Med. **191**(4), 603–612 (2000)
5. Lai, T.-S., Slaughter, T.F., Peoples, K.A., Hettasch, J.M., Greenberg, C.S.: Regulation of human tissue transglutaminase function by magnesium-nucleotide complexes Identification of distinct binding sites for Mg-GTP and Mg-ATP. J. Biol. Chem. **273**(3), 1776–1781 (1998)
6. Marrano, C., de Macédo, P., Gagnon, P., Lapierre, D., Gravel, C., Keillor, J.W.: Synthesis and evaluation of novel dipeptide-bound 1,2,4-thiadiazoles as irreversible inhibitors of guinea pig liver transglutaminase. Bioorg. Med. Chem. **9**(12), 3231–3241 (2001)

7. Mangala, L.S., Mehta, K.: Tissue transglutaminase (TG2) in cancer biology. Prog. Exp. Tumor Res. **38**, 125–138 (2005)
8. Zemskov, E.A., Janiak, A., Hang, J., Waghray, A., Belkin, A.M.: The role of tissue transglutaminase in cell-matrix interactions. Front Biosci. **11**, 1057–1076 (2006)
9. Korponay-Szabó, I.R., Dahlbom, I., Laurila, K., Koskinen, S., Woolley, N., Partanen, J., Kovács, J.B., Mäki, M., Hansson, T.: Elevation of IgG antibodies against tissue transglutaminase as a diagnostic tool for coeliac disease in selective IgA deficiency. Gut **52**(11), 1567–1571 (2003)
10. Szondy, Z., Sarang, Z., Molnár, P., Németh, T., Piacentini, M., Mastroberardino, P.G., Falasca, L., et al.: Transglutaminase 2-/- mice reveal a phagocytosis-associated crosstalk between macrophages and apoptotic cells. Proc. Natl. Acad. Sci. **100**(13), 7812–7817 (2003)
11. Johnson, K.A., Rose, D.M., Terkeltaub, R.A.: Factor XIIIA mobilizes transglutaminase 2 to induce chondrocyte hypertrophic differentiation. J. Cell Sci. **121**(13), 2256–2264 (2006)
12. Park, D., Choi, S.S., Ha, K.-S.: Transglutaminase 2: a multi-functional protein in multiple subcellular compartments. Amino Acids **39**(3), 619–631 (2010)
13. Madagi, S.B., Parvatikar, P.P.: Sequence analysis and structural characterization of tissue transglutaminase 2(TG2) by *In silico* approach. Int. J. Pharm. Pharmaceut. Sci. **9**(10), 37–42 (2017)
14. Yuan, L., Choi, K., Khosla, C., Zheng, X., Higashikubo, R., Chicoine, M.R., Rich, K.M.: Tissue transglutaminase 2 inhibition promotes cell death and chemosensitivity in glioblastomas. Mol. Cancer Ther. **4**(9), 1293–1302 (2005)
15. Verma, A., Wang, H., Manavathi, B., Fok, J.Y., Mann, A.P., Kumar, R., Mehta, K.: Increased expression of tissue transglutaminase in pancreatic ductal adenocarcinoma and its implications in drug resistance and metastasis. Can. Res. **66**(21), 10525–10533 (2006)
16. Jones, R.A., Kotsakis, P., Johnson, T.S., Chau, D.Y.S., Ali, S., Melino, G., Griffin, M.: Matrix changes induced by transglutaminase 2 lead to inhibition of angiogenesis and tumor growth. Cell Death Differ. **13**(9), 1442 (2006)
17. Herman, J.F., Mangala, L.S., Mehta, K.: Implications of increased tissue transglutaminase (TG2) expression in drug-resistant breast cancer (MCF-7) cells. Oncogene **25**(21), 3049 (2006)
18. Choi, K., Siegel, M., Piper, J.L., Yuan, L., Cho, E., Strnad, P., Omary, B., Rich, K.M., Khosla, C.: Chemistry and biology of dihydroisoxazole derivatives: selective inhibitors of human transglutaminase 2. Chem. Biol. **12**(4), 469–475 (2005)
19. Ren, W., Qiao, Z., Wang, H., Zhu, L., Zhang, L.: Flavonoids: promising anticancer agents. Med. Res. Rev. **23**(4), 519–534 (2003)
20. Benavente-Garcia, O., Castillo, J.: Update on uses and properties of citrus flavonoids: new findings in anticancer, cardiovascular, and anti-inflammatory activity. J. Agric. Food Chem. **56**(15), 6185–6205 (2008)
21. Huang, M., Jin-Jian, L., Huang, M.-Q., Bao, J.-L., Chen, X.-P., Wang, Y.-T.: Terpenoids: natural products for cancer therapy. Expert Opin. Investig. Drugs **21**(12), 1801–1818 (2012)
22. Suffness, M., Cordell, G.A.: Antitumor alkaloids. Alkaloids Chem. Pharmacol. **25**, 1–355 (1985)
23. Madagi, S.B., Parvatikar, P.P.: Docking studies on phytochemical derivatives as tissue transglutaminase-2 (TG2) inhibitors against lung cancer. In: Proceedings of The World Congress on Engineering and Computer Science. Lecture Notes in Engineering and Computer Science, 23–25 October 2018, San Francisco, USA, pp. 69-72 (2018)
24. Mangal, M., et al.: NPACT: naturally occurring plant-based anti-cancer compound-activity-target database. Nucleic Acids Res. **41**, D1124–D1129 (2012)
25. Morris, G.M., Huey, R., Lindstrom, W., Sanner, M.F., Belew, R.K., Goodsell, D.S., Olson, A.J.: AutoDock4 and AutoDockTools4: automated docking with selective receptor flexibility. J. Comput. Chem. **30**(16), 2785–2791 (2009)

Deposition of Nanostructured TiO_2/NiO Heterojunction Solar Cells Using Spray Pyrolysis

Kingsley O. Ukoba(✉) and Freddie L. Inambao

University of KwaZulu-Natal, 238 Mazisi Kunene Rd, Glenwood, Durban 4041, South Africa
ukobaking@yahoo.com, inambaof@ukzn.ac.za

Abstract. This study looked at the deposition of nanostructured TiO_2/NiO using a solution processed deposition technique of spray pyrolysis. The different techniques for depositing metal oxide solar cells are discussed with emphasis on the spray pyrolysis. The spray pyrolysis was used to deposit nanostructured TiO_2/NiO heterojunction solar cells at 350 °C using conducting Indium tin oxide substrate. X-ray diffraction shows that the heterojunctions have a polycrystalline cubic structure with a preferred orientation along the (1 1 1) and (2 0 0) planes. The elemental properties show the presence of TiO_2 and NiO. The optical band gap, and other optoelectronic properties were also investigated. These findings will enhance the study of cheap, efficient and sustainable alternate materials for solar energy development and affordable energy in developing countries.

Keywords: Affordable · Developing countries · Nanostructured · Optical properties · Spray pyrolysis technique · Solar cells · TiO_2/NiO

1 Introduction

A major breakthrough in solar cell fabrication would be the ability to have large-scale production at an affordable cost [1]. The major obstacle in using silicon solar cells is the expensive nature of the material and the complexities involved in fabricating the solar cells [2]. Apart from the expensive and commercially available silicon-based solar cells, there are difficulties in scaling up existing methods of solar cell fabrication. Most of the available methods for deposition of solar cells require a stable and steady supply of electricity. This has discouraged manufacturers in developing countries due to their erratic power supply [3].

The solution to such electricity woes may be found in nanostructured metal oxides [4], due to the low cost of processing and the simplicity of deposition methods of nanostructured metal oxides. Nickel oxide (NiO) holds great promise being a p-type metal oxide with a vast range of applications [5–7]. Several methods have been used to deposit NiO with a view to optimising it for various applications. The deposition methods include sputtering, hydrothermal growth, sol-gel, and laser ablation [8–11]. However, the spray pyrolysis technique (SPT) is preferred for films because it allows coatings on large areas

S.-I. Ao et al. (Eds.): WCECS 2018, *Transactions on Engineering Technologies*, pp. 112–123, 2020.
https://doi.org/10.1007/978-981-15-6848-0_10

in thin layers with uniform thickness [12, 13]. SPT's simplicity, affordability and the possibilities for mass production singled it out for this study. It requires electricity only during deposition, which can be less than four hours per deposition.

SPT finds application in deposition of metal oxides owing to the simplicity of the technique, low equipment cost, low maintenance, and low power consumption. SPT does need electricity for storage of the SPT equipment. The quality and properties of the deposited films depend largely on the process parameters. The substrate surface temperature affects the output of the films. Higher substrate temperatures produce rougher and porous film, but lower temperatures produce cracked film. Deposition temperature also influences the crystallinity, texture, and other physical properties of deposited film [14]. Precursor solution affects morphology and properties of deposited film [15]. SPT is grouped into four processes by means of reaction type [16]. Process 1 involves the droplet residing on the surface as the solvent evaporates thereby making the solid react when dry. In Process 2, the solvent evaporates just before the droplet makes contact with the surface; dry solid impinges on it allowing for decomposition. Process 3 is known as true chemical vapour deposition. In this process, solvent vaporizes as the droplet approaches the substrate which means the solid melts and vaporizes. Thereafter, the vapour diffuses to the substrate to undergo heterogeneous reaction. Process 4 occurs in the vapour state.

This study investigated the deposition of a nanostructured metal oxide for possible use in fabrication of affordable and efficient heterojunction solar cells. The optical properties of a metal oxide play a vital role in its usage in the fabrication of optoelectronic devices [17]. The optical properties reveal information relating to the microscopic behaviour of the material.

1.1 Thin Film Solar Cells and Metal Oxide Solar Cells

Thin-film solar cells are the basis for nearly all the presently available commercial solar cells. Thin film solar cells attempt to bridge the cost disadvantages of silicon wafers by using thin films of semiconductors that are usually deposited onto a supporting substrate. The active layers are usually thin yet able to absorb large amounts of incident solar radiation due to the material's strong absorption capacity [18].

Thin-film solar cells have been deposited using a variety of techniques onto different substrates (metal or non-metal, flexible or rigid) to produce a variety of layers (absorber, contact, buffer, anti-reflection) [19]. This flexibility makes it easy for thin films and engineering of layers to be fine-tuned in order to improve device performance. Thin film solar cells can be further developed to produce a tandem-structure approach called the integrated-tandem-Solar-Cell (ITSC) system. Metal oxides are a highly sought after candidate for a variety of technological applications due to their special and tuneable properties. They have excellent optical, optoelectronic, magnetic, mechanical, and thermal properties, among other properties [20].

Nickel oxide (NiO) is one of the least researched metal oxides yet with promising properties in optoelectronic applications. It can be obtained as black or green crystalline powder having a density of 6.67 g/cm^3 and melting point of 1955 °C with chemical composition of nickel being 78.55% and oxygen 21.40% with a molar mass of 74.6928 g/mol, magnetic susceptibility of $+660.0 \cdot 10\,cm^3/mol$ to 6 cm^3/mol and refractive index of 2.1818 [21]. Nickel oxides exist in various oxidation states viz nickel trioxide or sesquioxide (Ni_2O_3), nickelous oxide (NiO), nickel dioxide (NiO_2), nickelosic oxide (Ni_3O_4) and nickel peroxide (NiO_4) [22]. Amongst these, NiO has rhombohedral or cubic structure and possesses a pale green colour. The stoichiometry of NiO is roughly indicated by the colour of the sample [23]. NiO is a p-type semiconductor having wide band gap from 3.5 eV to 4.0 eV [24] with excellent chemical stability. It also has excellent durability and electrochemical stability with a large range of optical densities owing to good optical, electrical and magnetic properties. A versatile material with multiple applications in: solid oxide fuel cells (nickel cermet), lithium ion micro batteries (lithium nickel oxide cathode), electrochromic coating, in aerospace as a light weight structural component, as a catalyst, as anti-ferromagnetic layers, as adhesive, p-type transparent conductive films, materials for gas or temperature sensors such as CO sensors, H_2 sensor, and formaldehyde sensors and the p-type layer for UV detectors [5–7, 25]. Also, sintered nickel oxide is used for producing nickel steel alloys which won the 1920 Nobel Prize in Physics for Charles Guillaume. NiO is used in the Edison battery (nickel-iron battery) and is used to make NiCd rechargeable batteries which were used in many electronic devices until the advent of the NiMH battery.

2 Methodology

2.1 Deposition

The chemicals used were of analytical reagent grade and were used without further purification. Distilled and deionized pure water was used during the course of the experiment. The solar cell was fabricated using a modified SPT as reported by Ukoba et al. [26]. Figure 1 is a pictorial representation of the experimental set-up. Prior to sample preparation, the glass and the indium tin oxide (ITO) coated glass used as substrates were cleaned ultrasonically as reported by Adeoye and Salau [27]. The titanium oxide (TiO_2) nanostructure thin film was prepared by mixing 3 mL of titanium ethoxide with 30 mL of distilled water and ethanol mixture and three droplets of acetic acid.

This was stirred for one hour before spraying on cleaned indium tin oxide (ITO) coated glass substrates maintained at about 350 °C. Deposition parameters such as substrate temperature, carrier gas flow rate and pressure were optimised to obtain quality films. The nanostructured nickel oxide (NiO) was deposited on the prepared ITO/TiO_2 layers using SPT. The precursor for NiO was obtained by preparing 0.05 M nickel acetate tetrahydrate in double distilled water. The precursors were thoroughly stirred for several minutes prior to spraying onto preheated substrates maintained at about 350 °C. Other deposition parameters were maintained to obtain good quality thin films.

2.2 Testing

The TiO_2 and NiO prepared on ITO were used to study the elemental, morphological, structural and other optoelectronic characteristics of TiO_2 and NiO using Energy Dispersive X-ray Spectrometer (EDS or EDX: "AZTEC OXFORD DETECTOR"), a ZEISS ULTRA PLUS Field Emission Gun Scanning Electron Microscope (FEGSEM), and a

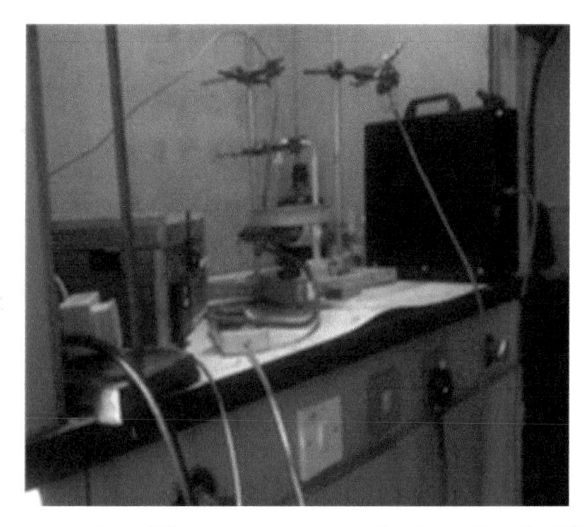

Fig. 1. Pictorial representation of the experiment set-up for depositing the TiO_2/NiO heterojunction solar cells

BRUKER AXS with D8 Advance diffractometer Cu-K α radiation X-ray Diffractometer (XRD). The weight difference method was also used. Optical properties were studied in the wavelength range of 300 nm to 1000 nm with a SHIMADZU UV-3600UV-VIS Spectrometer. The results of the characterised nanostructured TiO_2/NiO heterojunction solar cells are here reported.

3 Results and Discussion

3.1 Morphological Studies

Figures 2a and 2b show the scanning electron micrograph of the heterojunction of p-NiO/n-TiO_2 and the side view respectively.

The micrograph reveals scattered distribution and broader flakes of the TiO_2/NiO particles across the surface of the film. The film has even distribution, is adherent to the film surface, and has no cracks. This represents a better surface morphology compared to that of NiO films reported by Sriram and Thayumanavan [28]. This micrograph was obtained by the SEM at the point of interaction between the TiO_2 and NiO. It shows a polycrystalline structure. The micrograph shows the P-type NiO and N-type TiO2 of the thin film with their polycrystalline structures. It shows complete penetration at the heterojunction.

3.2 Elemental Composition

Figure 3 shows the elemental composition of the TiO_2/NiO heterojunction solar cell deposited on the ITO coated glass substrate. The figure shows the presence of Ti, O, and Ni in the TiO_2 and NiO respectively. This confirms the presence of the metal oxides in the heterojunction.

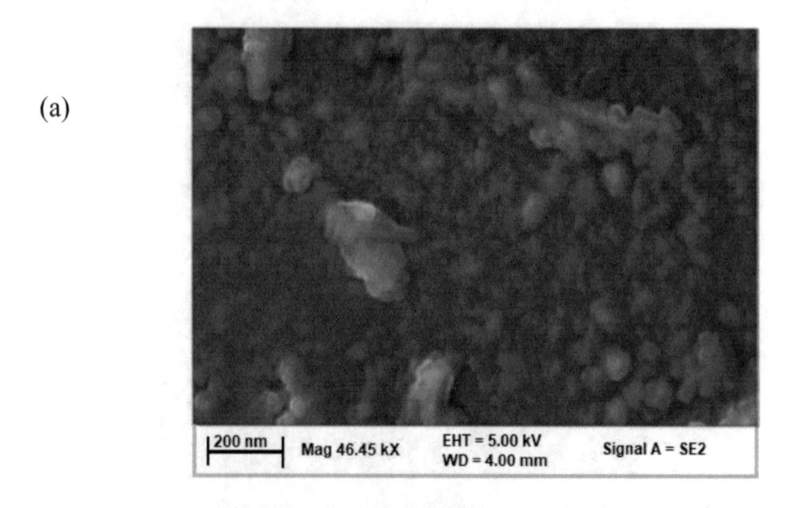

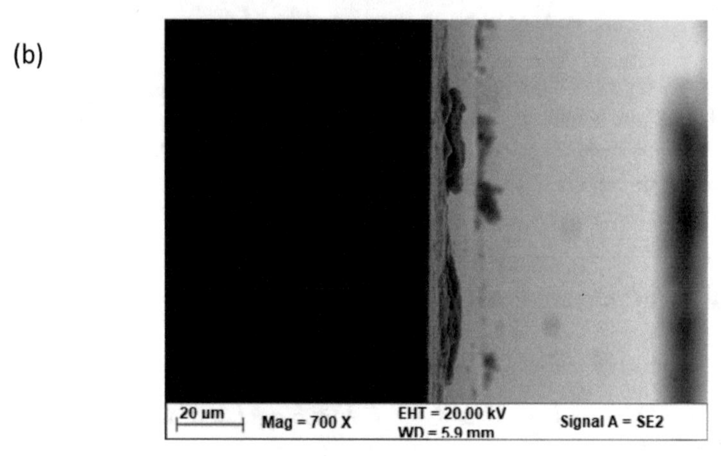

Fig. 2. SEM of the (a) TiO2/NiO heterojunction solar cell and (b) side view of the TiO2/NiO heterojunction solar cell

3.3 Structural Analysis

Figure 4 shows the X-ray diffraction patterns of the fabricated TiO_2/NiO heterojunction solar cell on the ITO substrate. The peaks corresponding to NiO and TiO_2 were determined with JCPDS patterns. The XRD spectrum indicates strong NiO peaks with (1 1 1), (2 0 0) and (2 2 0) preferential orientations. The patterns of the NiO thin film has peak diffractions at ($2\theta = 37°$, $2\theta = 43°$ and $2\theta = 64°$) for the (1 1 1), (2 0 0) and (2 2 0) planes. The XRD analysis confirms bunsenite which corresponds to the JCPDS card 04-0835 for nickel oxide [29] confirming it as a good absorber layer of solar cells [30]. The TiO_2 spectrum also shows strong spectrum and polycrystalline structures typical of N-type in heterojunction solar cells. The structure of the heterojunction indicates that the film is polycrystalline and chemically pure.

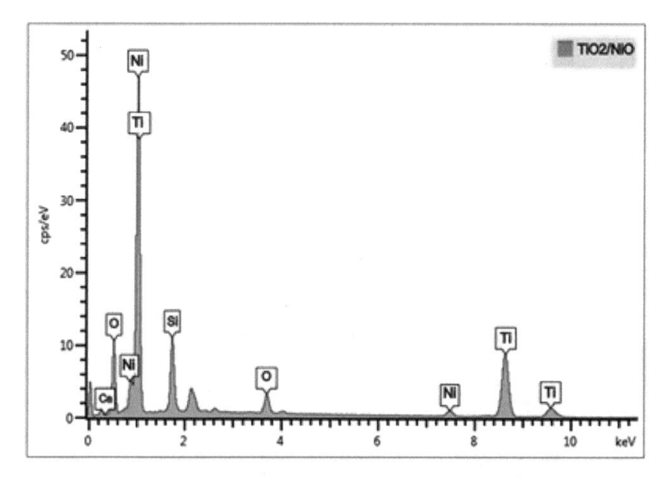

Fig. 3. Elemental composition of TiO_2/NiO

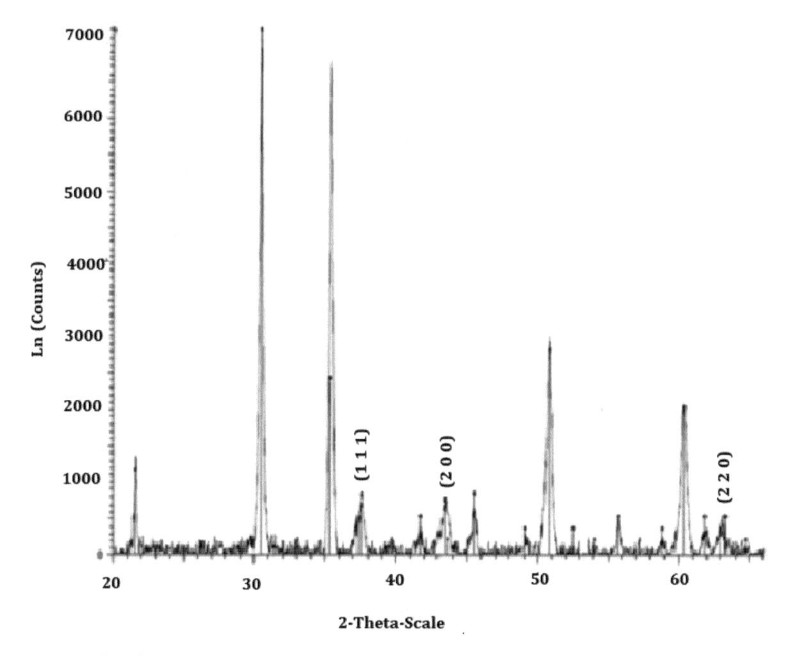

Fig. 4. Diffraction patterns of TiO_2/NiO heterojunction solar cells

3.4 Optical Properties

The optical properties of a semiconducting material are vital for understanding the electrical properties [31]. Transmittance and reflectance are required for measuring the absorption coefficient. The absorption coefficient measurement gives information about the energy band gap. The energy band gap is a major determinant of the electrical properties of the semiconductor.

3.5 Film Thickness

The measured data of the film thickness of the heterojunction is depicted in Fig. 5. The measured film thickness was found to be 4.39 μm. This is due to the joint deposition of TiO_2 and NiO on the ITO.

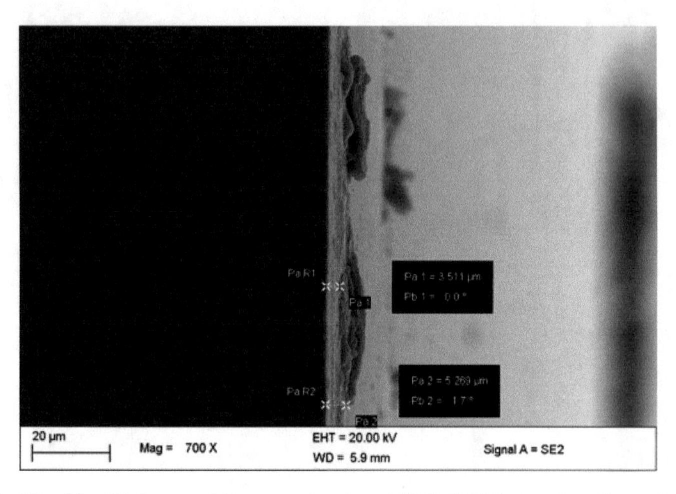

Fig. 5. The film thickness of the nanostructured TiO_o/NiO heterojunction solar cells

3.6 Transmittance

Figure 6 represents the measurement of transmittance for deposited nanostructured TiO_2/NiO heterojunction solar cells. The graph shows a control graph of TiO_2/NiO deposited on glass and the nanostructured TiO_2/NiO deposited on ITO substrate. The nanostructured TiO_2/NiO has a better transmittance of 91% when compared to that on a glass substrate of 58%. The nanostructured TiO_2/NiO deposited on a glass substrate is denser than the one on ITO substrate. The absorption peak is shifted to lower energies.

3.7 Absorbance

Figure 7 shows the absorbance of the TiO_2/NiO heterojunction on ITO and a glass substrate. This shows marked resemblance to the standard absorbance characteristic of transition metals (to which Ti and Ni belong) [32].

The heterojunction on ITO substrates absorbs more compared to the heterojunction deposited on the glass substrate. An indication of better performance and usage in solar cell fabrication [33].

3.7.1 Absorption Coefficient (A)

Absorption coefficient, α was obtained using Eq. (1) [34]:

$$\alpha = (2.303 \times A)/t \tag{1}$$

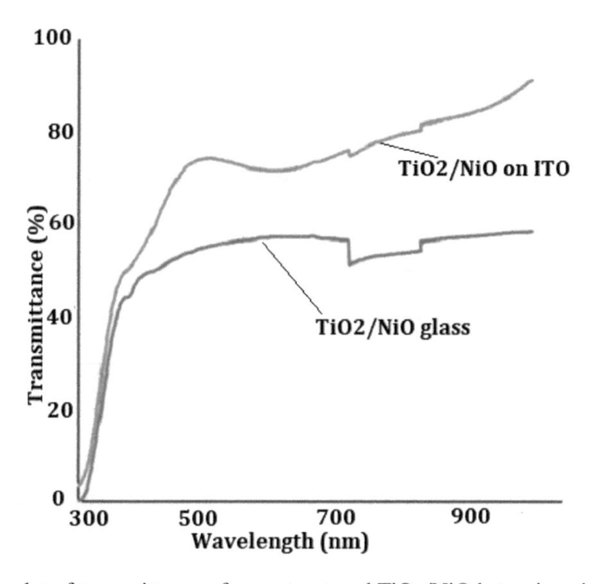

Fig. 6. The plot of transmittance of nanostructured TiO_2/NiO heterojunction solar cells

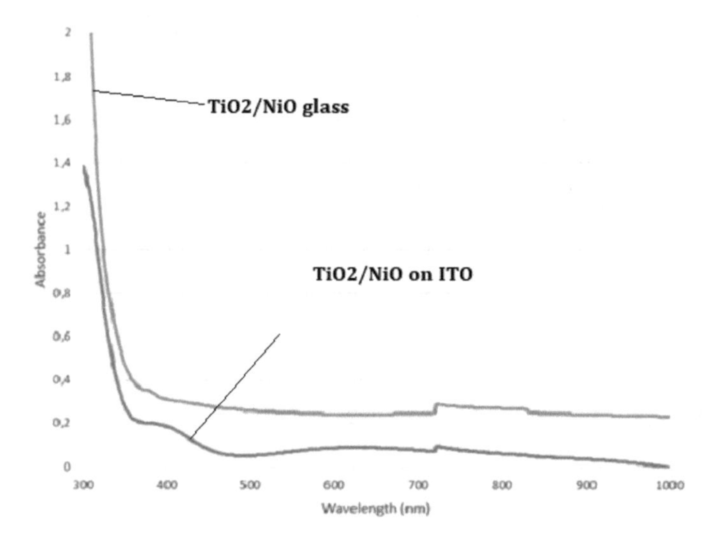

Fig. 7. The plot of absorbance of nanostructured TiO_2/NiO heterojunction solar cells

Where t is film thickness and A is absorbance.

The absorption coefficient is a vital parameter in the determination of the optical band gap. The absorption coefficient of the nanostructured TiO_2/NiO heterojunction solar cell is shown in Fig. 8.

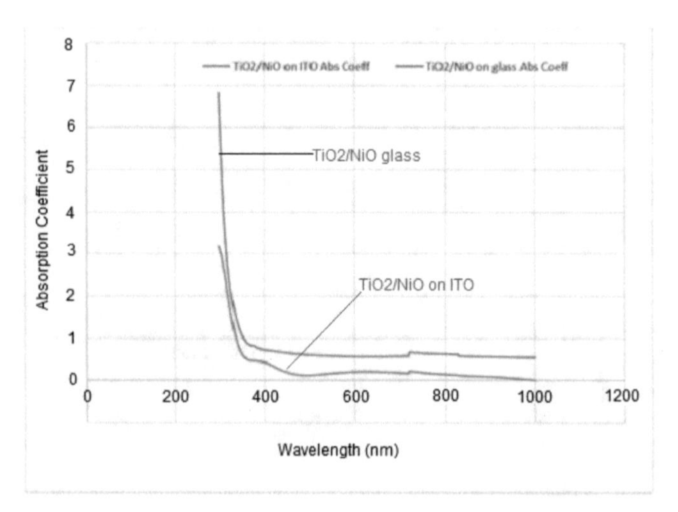

Fig. 8. The plot of the absorption coefficient of nanostructured TiO_2/NiO heterojunction solar cell

3.7.2 Optical Band Gap

According to Asogwa et al. [35] and Extrella et al. [36], the relationship between optical absorption and optical energy band gap is expressed in Eq. (2):

$$\alpha^2 = C\left(h\upsilon - E_g\right) \tag{2}$$

Where C has a constant value, h denotes Planck's constant, υ represent incidence light frequency, and E_g denotes optical energy band gap.

Figure 9 shows a graph of $(\alpha h\upsilon)^2$ against $h\upsilon$ for TiO_2/NiO heterojunction deposited on ITO and glass substrate. Extrapolation of Fig. 9 to the $h\upsilon$ axis for $(\alpha h\upsilon)^2 = 0$ gives the optical band gap. A shift towards lower energy is observed for optical band gap value. The reduction is attributed to the Moss-Burstein shift [37]. Optical energy band gaps are 3.67 eV and 3.875 eV for ITO and glass substrate respectively. The ITO substrate values compare favourably with the optical band gap value of 3.5 eV reported by Boschloo and Hagfeldt [24]. The quantum size effect may be responsible for the large value of the band gap [2]. Careful and well-optimised deposition parameters also help in obtaining a better optical band gap.

A positive gradient of 0.9419 was obtained as seen from the generated equation shown in Eqs. (3) and (4).

$$y = 0.9419x - 1.6816 \tag{3}$$

$$R^2 = 0.2912 \tag{4}$$

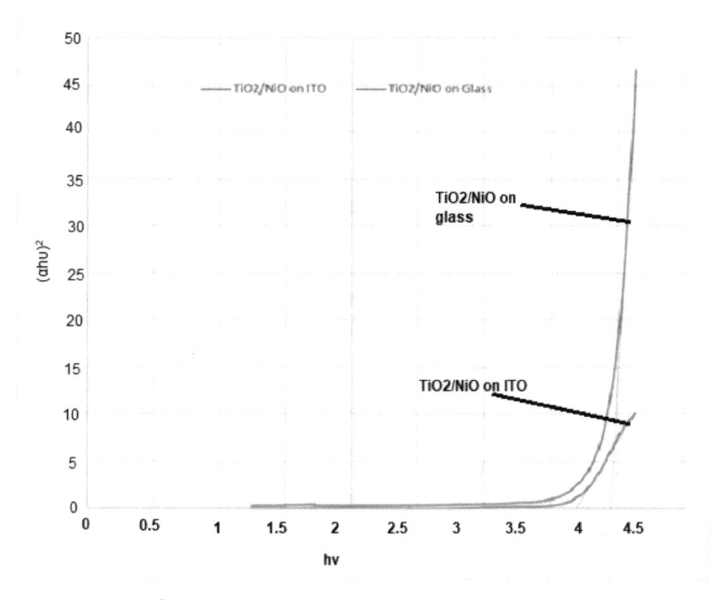

Fig. 9. Graph of $(\alpha h\upsilon)^2$ against $h\upsilon$ for nanostructured TiO_2/NiO heterojunction solar cells

4 Conclusion

This study investigated the optoelectronic properties of nanostructured TiO2/NiO heterojunction solar cells deposited on ITO substrates at 350 °C. The surface morphology shows the P-type NiO and N-type TiO_2 of the thin film with their polycrystalline structures. Both the TiO_2 and NiO had complete penetration at the heterojunction. The XRD spectrum indicates strong NiO peaks with (1 1 1), (2 0 0) and (2 2 0) at preferential orientation at $2\theta = 37°$, $2\theta = 43°$ and $2\theta = 64°$ respectively. Film thickness was found to be 4.39 μm. The nanostructured TiO_2/NiO have a better transmittance of 91%. Optical energy band gap was 3.67 eV. A positive gradient of 0.9419 was obtained for the TiO_2/NiO heterojunction deposited on ITO substrate. Extinction coefficients of TiO_2/NiO heterojunction deposited on ITO substrate vary within the UV region and are almost constant for both visible and near-infrared regions. The refractive index was found to be 1.94. This improved heterojunction is recommended for affordable and efficient solar cell fabrication especially in developing countries.

Acknowledgement. This work was supported by the National Research Foundation and The World Academy of Science (NRF-TWAS) of South Africa under grant number 105492.

References

1. Eslamian, M.: Spray-on thin film PV solar cells: advances, potentials and challenges. Coatings **4**(1), 60–84 (2014)

2. Ukoba, K.O., Inambao, F.L.: Study of optoelectronic properties of nanostructured TiO_2/NiO heterojunction solar cells. In: Proceedings of the World Congress on Engineering and Computer Science 2018, WCECS 2018, pp. 204–209, San Francisco, USA, 23–25 October 2018 (2018)
3. IRENA. Solar PV in Africa: Costs and Markets (2016). http://sun-connectnews.org/fileadmin/DATEIEN/Dateien/New/IRENA_Solar_PV_Costs_Africa_2016.pdf
4. Serrano, E., Rus, G., Garcia-Martinez, J.: Nanotechnology for sustainable energy. Renew. Sustain. Energy Rev. **13**(9), 2373–2384 (2009)
5. Nam, W.J., Gray, Z., Stayancho, J., Plotnikov, V., Kwon, D., Waggoner, S., Shenai-Khatkhate, D.V., Pickering, M., Okano, T., Compaan, A., Fonash, S.J.: ALD NiO thin films as a hole transport-electron blocking layer material for photo-detector and solar cell devices. ECS Trans. **66**(1), 275–279 (2015)
6. Wu, C.-C., Yang, C.-F.: Effect of annealing temperature on the characteristics of the modified spray deposited Li-doped NiO films and their applications in transparent heterojunction diode. Sol. Energy Mater. Sol. Cells **132**, 492–498 (2015)
7. Zhu, Z., Bai, Y., Zhang, T., Liu, Z., Long, X., Wei, Z., Wang, Z., Zhang, L., Wang, J., Yan, F., Yang, S.: High-performance hole-extraction layer of sol-gel-processed NiO nanocrystals for inverted planar perovskite solar cells. Angew. Chem. **126**(46), 12779–12783 (2014)
8. Hussein, M.M.A.: Optical and structural characteristics of NiO thin films doped with AgNPs by sputtering method. INAE Lett. **2**(1), 35–39 (2017)
9. Kerli, S., Alver, Ü.: Preparation and characterisation of ZnO/NiO nanocomposite particles for solar cell applications. J. Nanotechnol. (2016). https://doi.org/10.1155/2016/4028062
10. Sasi, B., Gopchandran, K.: Preparation and characterization of nanostructured NiO thin films by reactive-pulsed laser ablation technique. Sol. Energy Mater. Sol. Cells **91**(15–16), 1505–1509 (2007)
11. Zrikem, K., Da Rocha, M., Aghzzaf, A.A., Amjoud, M.B., Mezzane, D., Rougier, A.: Optimization of NiO thin films by sol-gel for electrochromic properties. In: ECS Meeting Abstracts. Abstract MA2017-01 1866 (2017)
12. Gowthami, V., Perumal, P., Sivakumar, R., Sanjeeviraja, C.: Structural and optical studies on nickel oxide thin film prepared by nebulizer spray technique. Phys. B Condens. Matter **452**, 1–6 (2014)
13. Faraj, M.G.: Effect of aqueous solution molarity on the structural and electrical properties of spray pyrolysed lead sulfide (PbS) Thin Films. Int. Lett. Chem. Phys. Astron. **57**, 122 (2015)
14. Filipovic, L, Selberherr, S, Mutinati, G.C., Brunet, E., Steinhauer, S., Köck, A., Grogger, W.: Modeling and analysis of spray pyrolysis deposited SnO2 films for gas sensors. In: Yang, G.-C., et al. (eds.) Transactions on Engineering Technologies, pp. 295–310. Springer, Dortmund (2014)
15. Perednis, D., Gauckler, L.J.: Thin film deposition using spray pyrolysis. J. Electroceramics **14**(2), 103–111 (2005)
16. Viguie, J.C., Spitz, J.: Chemical vapor deposition at low temperatures. J. Electrochem. Soc. **122**(4), 585–588 (1975)
17. Gowthami, V., Meenakshi, M., Perumal, P., Sivakumar, R., Sanjeeviraja, C.: Optical dispersion characterization of NiO thin films prepared by nebulized spray technique. Int. J. ChemTech Res. **6**(13), 5196–5202 (2014)
18. Tang, Z., Tress, W., Inganäs, O.: Light trapping in thin film organic solar cells. Mater. Today **17**(8), 389–396 (2014)
19. Chopra, K.L., Paulson, P.D., Dutta, V.: Thin-film solar cells: an overview. Prog. Photovolt. Res. Appl. **12**(2–3), 69–92 (2004)
20. Ukoba, O.K., Eloka-Eboka, A.C., Inambao, F.L.: Review of nanostructured NiO thin film deposition using the spray pyrolysis technique. Renew. Sustain. Energy Rev. **82**(3), 2900–2915 (2018). https://doi.org/10.1016/j.rser.2017.10.041

21. Hussein, M.A.M.: Effect of magnetization on ferrromagnetic and antiferromagnetic for NiO properties using Quantum ESPRESSO Package. Doctoral dissertation, Sudan University of Science and Technology (2016)
22. Subramanian, B., Ibrahim, M.M., Senthilkumar, V., Murali, K.R., Vidhya, V.S., Sanjeeviraja, C., Jayachandran, M.: Optoelectronic and electrochemical properties of nickel oxide (NiO) films deposited by DC reactive magnetron sputtering. Phys. B Condens. Matter **403**(21–22), 4104–4110 (2008)
23. Kunz, A.B.: Electronic structure of NiO. J. Phys. C: Solid State Phys. **14**(16), L455 (1981)
24. Boschloo, G., Hagfeldt, A.: Spectroelectrochemistry of nanostructured NiO. J. Phys. Chem. B **105**(15), 3039–3044 (2001)
25. Park, N., Sun, K., Sun, Z., Jing, Y., Wang, D.: High efficiency NiO/ZnO heterojunction UV photodiode by sol–gel processing. J. Mater. Chem. C **1**(44), 7333–7338 (2013)
26. Ukoba, K., Inambao, F., Eloka-Eboka, A.: Fabrication of affordable and sustainable solar cells using NiO/TiO2 PN heterojunction. Int. J. Photoenergy **2018**, 7 (2018). https://doi.org/10.1155/2018/6062390
27. Adeoye Abiodun, E., Salau, A.: Effect of annealing on the structural and photovoltaic properties of cadmium sulphide: copper sulphide (Cds: Cuxs) heterojunction. Int. J. Sci. Res. Publ. **5**(8), 1–5 (2015)
28. Sriram, S., Thayumanavan, A.: Structural, optical and electrical properties of NiO thin films prepared by low cost spray pyrolysis technique. Int. J. Mater. Sci. Eng. **1**, 118–121 (2013)
29. Gabal, M.: Non-isothermal decomposition of NiC2O4–FeC2O4 mixture aiming at the production of NiFe2O4. J. Phys. Chem. Solids **64**(8), 1375–1385 (2003)
30. Bakr, N.A., Salman, S.A., Shano, A.M.: Effect of co doping on structural and optical properties of NiO thin films prepared by chemical spray pyrolysis method. Int. Lett. Chem. Phys. Astron. **41**, 15–30 (2015)
31. Benno, G., Joachim, K.: Optical properties of thin semiconductor films (2003). home.fnal.gov/~jkopp/tum/pdf/F/hl_spekt.pdf
32. Axelevitch, A., Gorenstein, B., Golan, G.: Investigation of optical transmission in thin metal films. Phys. Procedia **32**, 1–13 (2012)
33. Ezugwu, S., Ezema, F., Osuji, R., Asogwa, P., Ekwealor, A., Ezekoye, B.: Effect of deposition time on the band-gap and optical properties of chemical bath deposited CdNiS thin films. Optoelectron. Adv. Mater. Rapid Commun. **3**(2), 141–144 (2009)
34. Barman, J., Sarma, K., Sarma, M., Sarma, K.: Structural and optical studies of chemically prepared CdS nanocrystalline thin films. Indian J. Pure Appl. Phys. **46**, 339–343 (2008)
35. Asogwa, P., Ezugwu, S., Ezema, F., Ekwealor, A., Ezekoye, B., Osuji, R.: Effect of thermal annealing on the band gap and optical properties of chemical bath deposited PbS-CuS thin films. J. Optoelectron. Adv. Mater. **11**(7), 940–944 (2009)
36. Estrella, V., Nair, M., Nair, P.: Semiconducting Cu3BiS3 thin films formed by the solid-state reaction of CuS and bismuth thin films. Semicond. Sci. Technol. **18**(2), p190 (2003)
37. Moss, T.: The interpretation of the properties of indium antimonide. Proc. Phys. Soc. Sect. B **67**(10), 775 (1954)

Solid-State Protection of a Perturbed Electric Power System Network

Muncho Josephine Mbunwe(✉), Boniface Onyemaechi Anyaka, and Uche Chinweoke Ogbuefi

Department of Electrical Engineering, University of Nigeria Nsukka, Nsukka, Enugu State, Nigeria
mamajoesix@gmail.com, {muncho.mbunwe,boniface.anyaka, uche.ogbuefi}@unn.edu.ng

Abstract. With the ever-increasing complexity of disturbed Power Systems, the method of protection using electromagnetic relays may not be adequate to afford appropriate discrimination especially when the fault current flows in parallel paths. Because the electromagnetic relay has slow response time, and the fact that it is a mechanical device and has moving parts, over a period of time, these moving parts will wear out and the relay will fail. To overcome these drawbacks, a more reliable means of system protection was carried out using solid-state contactless devices. A modular technique using five blocks made up of power supply unit, current sensor unit, voltage sensor unit, switching unit and output unit were used in the study design. A 66 kV power network was used to test the reliability of the protection designed circuit. Result shows enhanced system security.

Keywords: Fault · Mechanical device · Opto-coupler · Power system · Protection · Solid-state relay

1 Introduction

Power system protection is an act of ensuring availability of electric power without any interruption to every load connected to it [1]. The main purpose of power system protection is to isolate a faulty section of electrical power system from rest of the live system so that the rest potion can function satisfactorily without any sever damage due to fault current [2]. Overload, temporary faults, permanent faults, lightning strokes, induced lightning surges are of common occurrences in power systems. It is there important to ensure that the fault part is quickly disconnected from rest of the system so that damage is minimum. Most common type of abnormal conditions and faults in power system are: over-currents (overloads), sparking and arcing-grounds on overhead lines, short-circuits, flashovers due to lightning surges, faults due to insulation failure, temporary over-voltages/under-voltages of power frequency, unbalance of three phase voltages, etc. The faults on overhead lines are either temporary/transient or permanent [2]. Protective devices consist of mainly power system relays like current relays, voltage relays, impedance relays, power relays, frequency relays, etc. During fault, the protective relay

S.-I. Ao et al. (Eds.): WCECS 2018, *Transactions on Engineering Technologies*, pp. 124–138, 2020.
https://doi.org/10.1007/978-981-15-6848-0_11

gives signal to the associated circuit breaker for opening its contact. All the circuit breakers of electric power systems are direct current (dc) operated. Because dc can be stored in battery and if situation comes when total failure of incoming power occurs, still the circuit breakers can be operated for restoring the situation by the power of storage station battery. Hence, the battery is another essential item of the power system. Auto-reclosing of circuit breakers is adopted for overhead power systems to achieve service continuity after temporary faults. Auto-reclosing circuit breakers are used for overhead distribution feeders. They are not used for underground systems [3].

1.1 Components of Protective Schemes

1) Current Transformers (CTs)
   In order to obtain currents which can be used in control circuits and that are proportional to the system primary currents, current transformers are used. Often the primary conductor itself, for example a bus-bar, forms a single primary turn (bar primary). Whereas instrument current transformers have to remain accurate only up to slight over-currents, protection current transformers must retain proportionality up to at least 20 times normal full load. The normal secondary current rating of current transformers is now usually 1 A, but 5 A has been used in the past. A major problem can exist when two current transformers are used which should retain identical characteristics up to the highest fault current, for example in pilot wire schemes. Because of saturation in the silicon steel used and the possible existence of a direct component in the fault current, the exact matching of such current transformers is difficult.
2) Linear Couplers
   The problem associated with current transformers have resulted in the development of devices called linear couplers, which serve the same purpose but, having air-cores, remain linear at the highest currents. These are also known as Rogowski coils and are particularly suited to digital protection schemes.
3) Relays
   A relay is a device which, when supplied with appropriately scaled quantities, indicates an abnormal or fault condition on the power system, when the relay contacts close, the associated circuit-breaker trip-circuits are energized and the breaker opens, isolating the faulty part of the power network. Historically electromagnetic and semiconductor relays were installed and are still in use on the system. Modern practice is to install digital (numerical) protection. Although now almost all new relays use micro-processors and the measured quantities are manipulated digitally, the underlying techniques are often those developed for electro-mechanical relays. The use of electromechanical relay (EMR) in protecting perturbed electric network has been in use in power system engineering over years. However, the EMR are inexpensive, easy to use and allow the switching of load circuit controlled by low power, electrically isolated input signal [4]. One of the main disadvantages of an EMR is that it is a "mechanical device", that is, it has moving parts so the switching speeds (response time) due to physical movement of the metal contacts using a magnetic field is slow. Over a period of time these moving parts will wear out and fail, or that the contact resistance through the constant arcing and erosion may make the relay unstable and

shortens its life span. Also, it is electrically noisy with the contacts suffering from contact bounce which may affect any electronic circuits to which it is connected [5]. However, as a result of these disadvantages, there is need therefore, to device a more reliable means to overcome these disadvantages, hence the use of a solid-state relay (SSR) which is a solid state contactless, pure electronic relay. The SSR being purely electronic devices has no moving parts within its design as the mechanical contacts have been replaced by power transistors, thyristors or triac. The electrical separation between the input control signal and the output load voltage is accomplished with the aid of an opto-coupler (a component that transfers electrical signals between two isolated circuits by using light) type light sensor. Continuous rise in interconnection of loads/systems to the power system topology of most countries has resulted in higher demand of power which in most cases cannot be provided or supplied by generating stations (with several generating units in a single area network). To meet up with the demand, in most cases it leads to faults occurring on the power system network, such as: over loading; short circuit; and over voltage and under voltage. It is as a result of these operational challenges encountered in the Nigeria Power System that motivated this work.

The objectives of this work are:

- To isolate a faulty section of electrical power system from the rest of the live system so that the portion can function satisfactorily without any severe damage due to fault current.
- To increase efficiency and reduce cost of protection by the use of SSR.
- To prevent damage to the system apparatus from hazards (like fire) with the help of SSR.
- To greatly improve on the transient state stability limit of the system.
- To avoid permanent damage to the equipment and also to reduce possibility of developing simplest fault into more severe fault.

2 System Protection

Power system protection emerged at the beginning of the last century, with the application of the first electro-mechanical overcurrent relay. The majority of the protection principles currently employed in protection relays were developed within the first three decades of the last century, such as over current, directional, distance and differential protection. The development of modern science and technology, especially electronic and computer technology, promoted the development of relay technology, such as materials, components and the manufacturing process of the hardware structure of relay protection device. At the same time, great theoretical progress had been made in the relay protection software, algorithms, etc. [6]. The progress in modern technology stimulates the development in power system protection. In the last century, relay protection had gone through a number of development stages, migrating from mechanical to electromechanical, and subsequently to solid state semiconductor technologies. Today, solid state relays are replacing conventional relays in all areas of power system protection.

However, many of the same relaying principles of protection are still playing dominant role to date [7]. In order to attain the desired reliability, the power system network is divided into different protection zones: generator protection, transformer protection, transmission line protection, bus protection and feeder protection [8].

2.1 Generator Protection

A generator is subjected to electrical traces imposed on the insulation of the machine, mechanical forces acting on the various parts of the machine, and temperature rises. Even when properly used, a machine in its perfect running condition does not only maintain its specified rated performance for many years, but it does also repeatedly withstand certain excess of over load. Hence, preventive measures must be taken against overloads and abnormal conditions of the machine so that it can serve safely. Despite of sound, efficient design, construction, operation, and preventive means of protection, the risk of that fault cannot be completely eliminated from any machine [8]. The devices used in generator protection, ensure the fault, is rectified as quickly as possible. An electrical generator can be subjected to either internal fault or external fault or both. The generators are normally connected to an electrical power system, hence any fault occurred in the power system should also be cleared from the generators as soon as possible otherwise it may create permanent damage in the generator. The number and variety of faults occur in generator are huge. Great care is to be taken in coordinating the systems used and the settings adopted, so that the sensitive, selective and discriminative generator protection scheme is achieved [9]. The various forms of protection applied to the generator can be categorized into two manners: protective relays to detect faults occurring outside the generator and protective relays to detect faults occurring inside the generator. Other than protective relays, associated directly with the generator and its associated transformer, there are lightening arrestors, over speed safe guards, oil flow devices and temperature measuring devices for shaft bearing, stator winding, transformer winding and transformer oil etc. some of these protective arrangements are of non-trip type which only generate alarm during abnormalities. But the other protective schemes ultimately operate master tripping relay of the generator. It should be noted that no protective relay can prevent fault, it only indicates and minimizes the duration of the fault to prevent high temperature rise in the generator otherwise there may be permanent damage. It is desirable to avoid undue stresses in the generator, and for that it is a usual practice to install surge capacitor or surge diverter or both to reduce the effects of lightning and other voltage surge on the machine.

2.2 Transformer Protection and Transformer Fault

There are different kinds of transformers for example: core or auto transformers etc. And in terms of windings there are: two-winding or three-winding transformers. Different transformers, demand different scheme of transformer protection depending on their importance such as: winding connections, earthing methods and mode of operation etc. It is a common practice to use Buchholz relay protection for all 0.5 MVA and above transformers. While for small size transformers, only high voltage fuses are used as main protective devices. For medium transformers, over current protection along with restricted earth fault protection is applied. Differential protection should be provided in the transformers rated above 5 MVA [10]. Although an electrical power transformer is a static device, but internal stresses arising from abnormal system conditions must be taken into consideration. Transformers generally suffer from the following types of faults: over current due to overloads and external short circuit, terminal fault, Winding fault, and incipient fault. These faults cause mechanical and thermal stresses inside the transformer winding and its connecting terminals. Thermal stresses lead to overheating which ultimately affect the insulation system of the transformer. Deterioration of insulation leads to winding faults. Sometime failure of transformer cooling system, leads to overheating of transformer. So, the transformer protection schemes are very much required. The short circuit current of a transformer is normally limited by its reactance and for low reactance; the value of short circuit current may be excessively high. Whatever may be the fault, the transformer must be isolated instantly during fault otherwise major breakdown may occur in the electrical power system.

2.3 Bus-Bar Protection/Bus-Bar Differential Protection Scheme

In the early days, only conventional over current relays were used for bus-bar protection. But it is desired that fault in any feeder or transformer connected to the bus-bar should not disturb bus-bar system. In viewing of this time setting of bus-bar protection relays are made lengthy. So, when faults occur on bus-bar, it takes much time to isolate the bus from the source which may cause much damage in the bus system [11]. In recent days, the second zone distance protection relays on incoming feeder with operating time of 0.3 to 0.5 s have been applied for bus-bar protection. The disadvantage of the scheme is that it cannot discriminate the faulty section of the bus-bar. Nowadays, electrical power system deals with huge amount of power. Hence any interruption in total bus system causes big loss to the company. So, it becomes essential to isolate only faulty section of bus-bar during bus fault. Another drawback of the second zone distance protection scheme is that sometimes the clearing time is not short enough to ensure the system stability. To overcome this difficulty, differential bus-bar protection scheme with an operating time less than 0.1 s, is commonly applied to high voltage bus systems using kirchhoff's current first law as shown in Fig. 1.

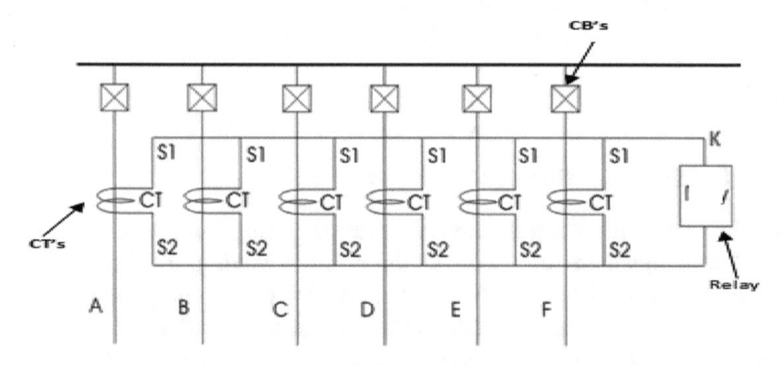

Fig. 1. Current differential protection circuit

In Fig. 1, the secondary of CTs are connected in parallel and S1 terminals of all CTs are connected together and a bus wire is formed. Similarly, S2 terminals of all CTs are connected together to form another bus wire. A tripping relay is connected across the two bus wires.

2.4 Protection of Transmission Lines and Feeders

Whenever the transmission line is long and run through open atmosphere, the probability of fault occurring in the transmission line is much higher than that of power transformers and alternators. That is why a transmission line requires much more protective schemes than transformers and alternators [12]. Protection of transmission line should have some special features such as:

- Tripping of circuit breaker closest to the fault point during fault.
- In case the circuit breaker closest is faulty, the next circuit breaker should trip as back up.
- The operating time of relay associated with protection of line should be as minimum as possible in order to prevent unnecessary tripping of circuit breakers association with other healthy parts of power system.

These requirements cause protection of transmission line much different from protection of transformer and other equipment of power systems [13]. There are three methods of protecting the transmission line: time graded over current protection, differential protection and distance protection.

2.5 Protective Relay

A relay can be operated with a small amount of power and can be used to control devices that need much more power like circuit breakers and isolators. A relay is like a remote-controlled switch and has many applications because of its long life, high accuracy, relative simplicity and proven high reliability [14]. This is very useful when controlling a requirement of huge amount of voltage or current with the use of a small electrical signal. In the industry, a wide variety of applications requires the use of relays. Electrical power systems can be protected against fault using sophisticated relays [15]. Initially, for a normally opened relay, the contact is normally open which means not connected. The coil generates a magnetic field when current (I) pass through it and closes the switch (i.e. top contact gets connected). A spring is used to pull back the switch open, when power is removed from the coil. Figure 2 shows the basic circuit for a normally opened relay operation while Fig. 3 shows the diagram of normally closed relay which operates in the opposite form of normally opened electromechanical relay [16].

Apart from electromechanical relays, modern power system network use SSR [17]. In this study three types of SSR are discussed: Reed Relay Coupled SSR showed in Fig. 4, Transformer Coupled SSR shown in Fig. 5, and Photo Coupled SSR shown in Fig. 6. The use of solid state protective relay type was adopted in this work due to its numerous advantages over other types of protective relays. These benefits are not limited to the following [18–20]

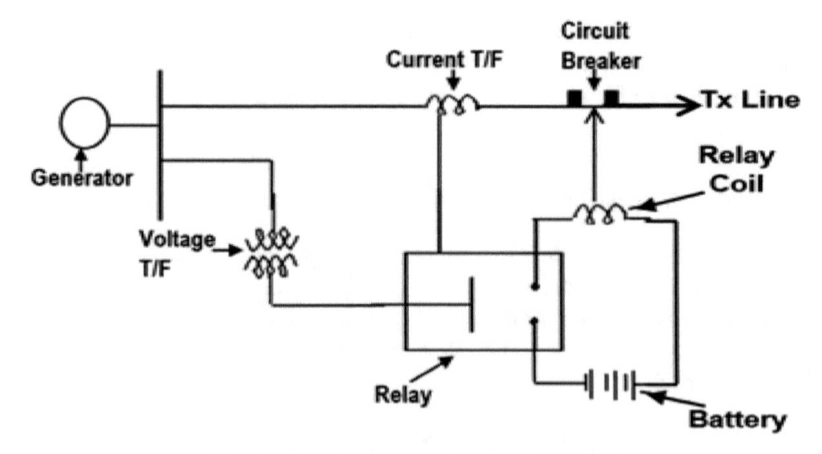

Fig. 2. Circuit diagram of normally opened

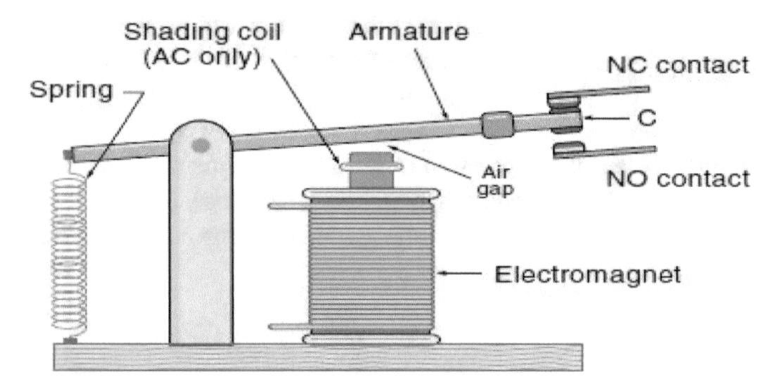

Fig. 3. Diagram of normally closed electromechanical relay.

- high degree of reliability;
- long operational life;
- zero-voltage turn-on, low electro-magnetic induction;
- shock and vibration resistant;
- no contact bounce which leads to arc less switching;
- microprocessor compatible;
- fast response;
- no effect of gravity or vibration or shock.

Sometimes, these relays use microprocessor but cannot be called microprocessor relays as it lacks the attribute of digital/numeric relay. These relays use semiconductor devices like diodes, SCR, TRIAC, Power transistor etc. to conduct load current. Relatively low control circuit energy is required to perform switching of the output state from OFF to ON position since semiconductor devices are used. To protect the circuit under control for introduction of electrical noises, the static relays are often used. Static relays are highly reliable and have a long life. It does not have any moving parts or contact bounce and thus have a fast response [18]. The classification of SSR by the nature of the input circuit is as follows [19].

a. Reed Relay coupled SSR application of control signal occurs on the coil of the reed relay. Thyristor switch is triggered when the appropriate circuitry is activated upon closing of reed switch as shown in Fig. 5.

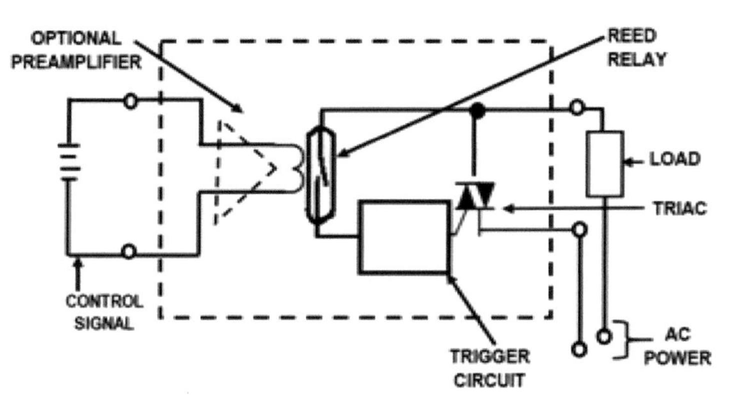

Fig. 4. Reed Relay Coupled SSR

b. Transformer coupled SSR has a low-power transformer primary consisting of the control signal. The thyristor switch is triggered by the secondary that is generated by the primary excitation as shown in Fig. 5.

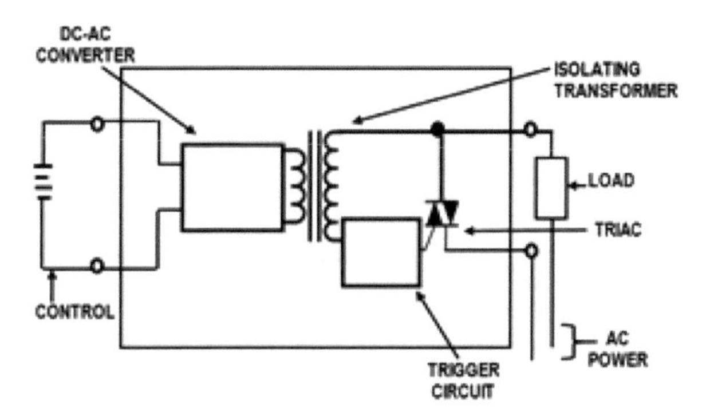

Fig. 5. Transformer Coupled SSR

c. Photo Coupled SSR as shown in Fig. 6, a light or infrared source (generally LED) consist of the control signal, photo-sensitive semi-conductor device (diode, transistor or thyristor) detects the radiation from that source and generates an output. This output triggers the TRIAC which is used to switch the load current. The electrical isolation is excellent as the input and output path are coupled only by a beam of light.

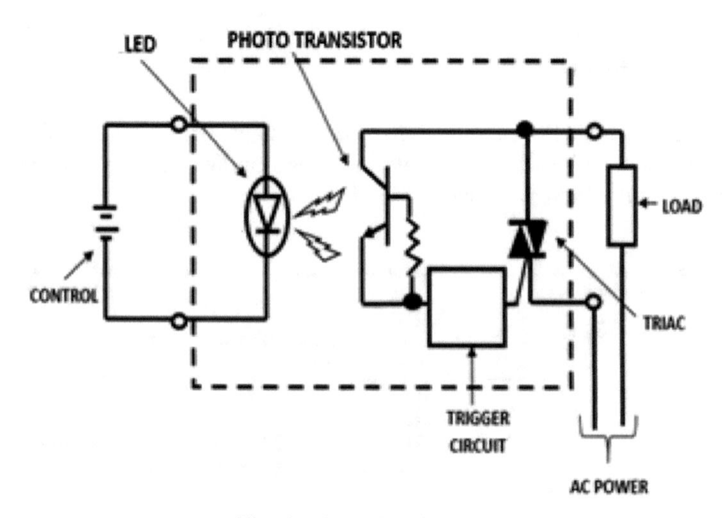

Fig. 6. Photo Coupled SSR

3 Design Method

This study has five distinct units which include the power supply unit, the voltage sensing unit, the current sensing unit, the switching unit and the output unit as shown in Fig. 7. The materials used to actualize this project include the following: dc power supply, diodes, capacitors, resistors, transistors, solid-state relay, operational amplifier (LM324), NOT gate, light emitting diodes (LEDs), voltage regulator, current transformer, lamp; variable resistors, and casing.

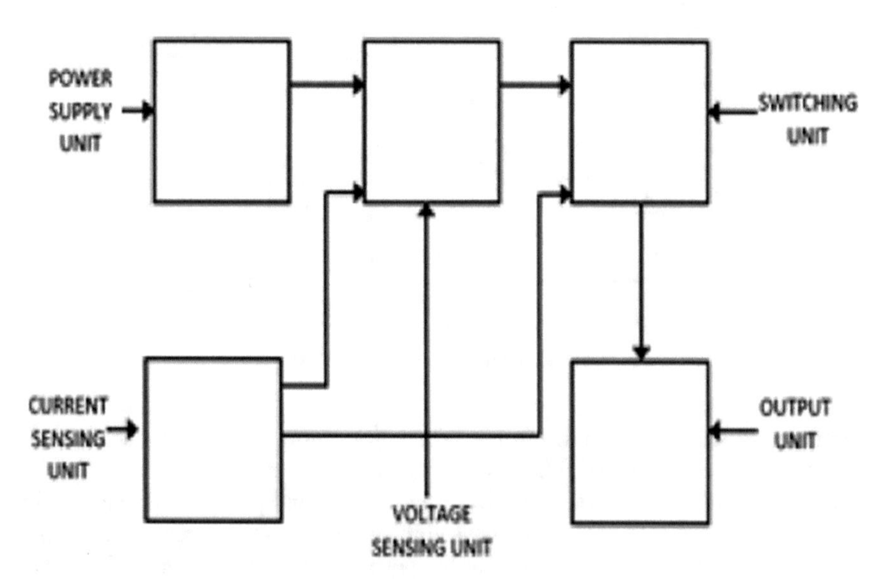

Fig. 7. Block diagram of the protection system using SSR.

3.1 The Power Supply Unit

This serves as input unit to the system. In this unit, power is converted from ac to dc and filtered as shown in Fig. 8.

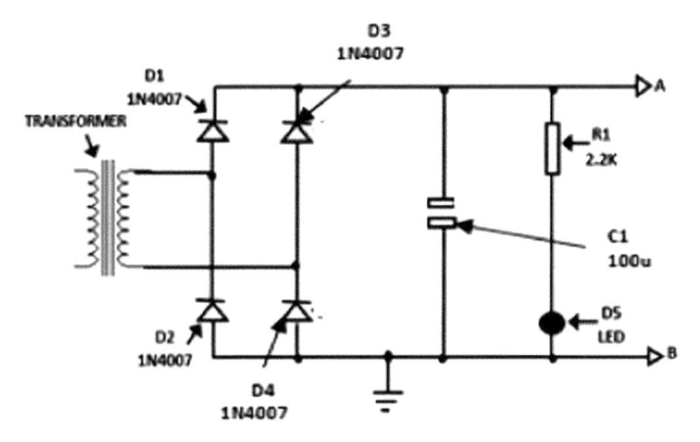

Fig. 8. Power supply unit.

3.2 The Voltage Sensing Unit

The voltage sensing unit shown in Fig. 9 consists of a voltage regulator (IC1), comparator (IC2), resistor (R3, R4, R5, R6 and R7), NOT gate and a light emitting diode. The circuit operates by stabilizing the rectified voltages 12 V dc with the help of voltage regulator. The resistors used to form the voltage dividers include (R3, R4, R6 and R7). The resistor R5, function as a limiting current resistor to the Not gate. The reduced voltages are compared through the inverting and the non-inverting inputs of the comparators. During the process, when the voltage in the non-inverting input is greater than the inverting, input the output of the comparator (IC2) becomes "high" (logic 1). But in a situation where the voltage in the inverting inputs is greater or equal to the non-inverting inputs, the output will become "low" (logic 0), and the LED in series with the NOT gate will come on indicating that there is no output from the comparator.

3.3 The Current Sensing Unit

The current sensing unit shown in Fig. 11 consists of current transformer, rectification diodes, filtering capacitors, resistor and a comparator (LM324N). The main function of this unit is to covert current to voltage. It monitors the current flowing through the conductor from the mains input to the output and then compares it with a pre-calibrated value (30 A). This unit monitors the current passing through the conductor to the output and coverts it into voltage, the ac voltage is then rectified with a full wave diode rectification using IN007. The direct current (dc) voltage is then divided using one fixed and one variable resistor, this is to make it possible to recalibrate the current sensor.

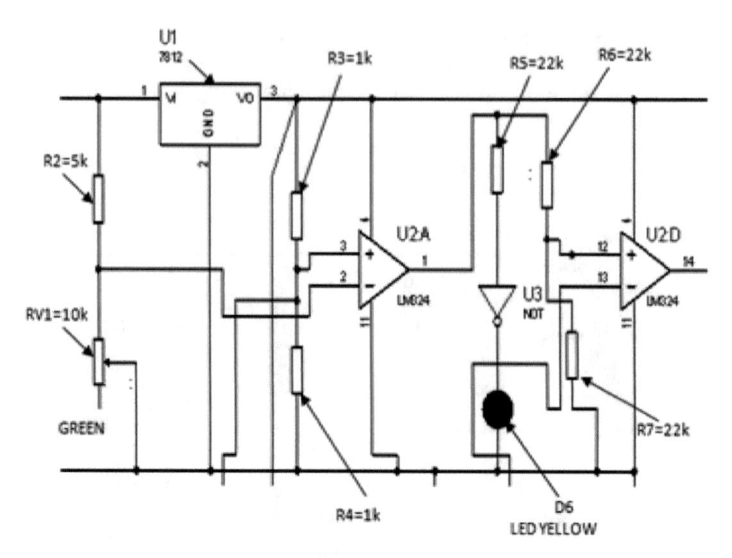

Fig. 9. Circuit diagram of the voltage sensing unit.

The output from the voltage divider is then passed through the non-inverting input of a comparator that has a reference voltage set to 6 V dc, when the non-inverting input of the comparator (LM 324 N) is greater than the inverting input, the system gives an output of 12 V dc.

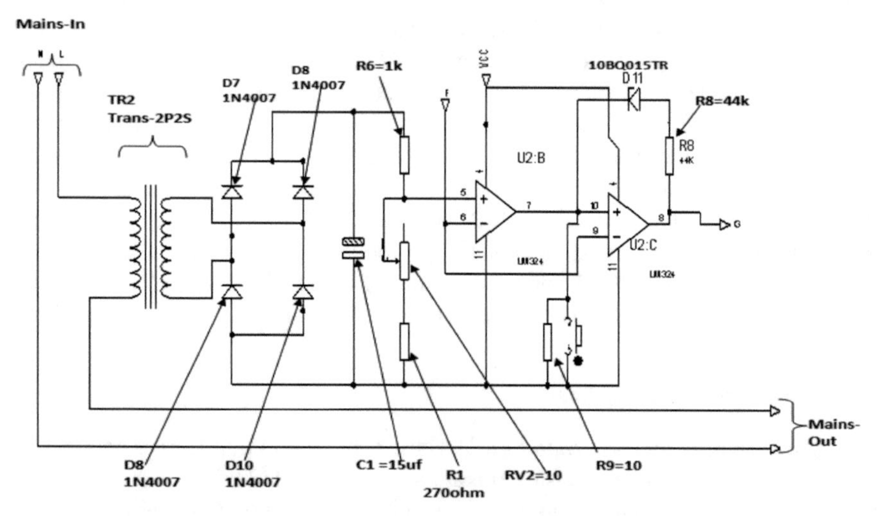

Fig. 10. Current sensing unit circuit diagram.

3.4 The Switching Unit

The switching unit of Fig. 11 is responsible for the making and breaking states of the relay. Transistors are used to drive the solid-state relays in this unit. The unit consists of an NPN transistor (Q1), resistors, and the solid-state relay. Transistor (Q1) conducts when its base senses voltage. This transistor is used to activate the collector current to the quantity required by the solid-state relay.

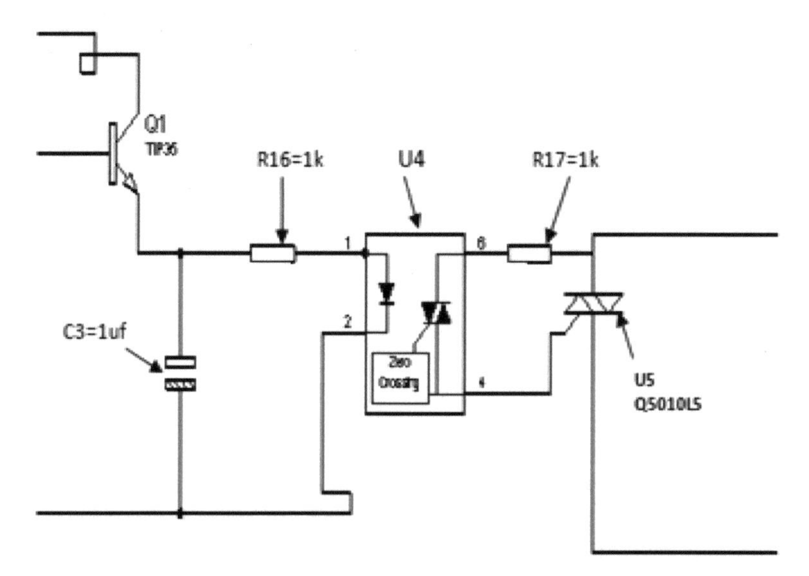

Fig. 11. Circuit diagram of the switching unit.

3.5 The Output Unit

The output unit of the protective system is used to determine the output state of the SSR during over voltage, under voltage, over current or short circuit fault conditions.

3.6 Designed System Analysis

The circuit diagram shown in Fig. 12 has four operational amplifiers that are configured as comparators. In Fig. 12, power is supplied to the V_{CC} (terminal 4) of the operational amplifier. The first operational amplifier (voltage sensing unit) compares voltages between pins 3 and 2 of the operational amplifier. The non-inverting terminal pin 3 is set to a reference voltage of $6V_{dc}$, terminal 1 gives output only when pin 3 is higher than pin 2. The 10 kΩ variable resistor in Fig. 9 is used to set the voltage between pin 2 and pin 3, such that pin 3 is higher than pin 2. The output signal from the comparator biases the second operational amplifier through 2.2 kΩ resistor which limits the flow of current to the second operational amplifier which is also configured to act as a comparator. The second operational amplifier compares the voltage between pin 12 and pin 13. Pin

12 acts as the reference voltage, the operational amplifier gives output when pin 12 is greater than pin 13. The output through pin 14 sends signal to the base of the transistor Q1 TIP36 of Fig. 11.

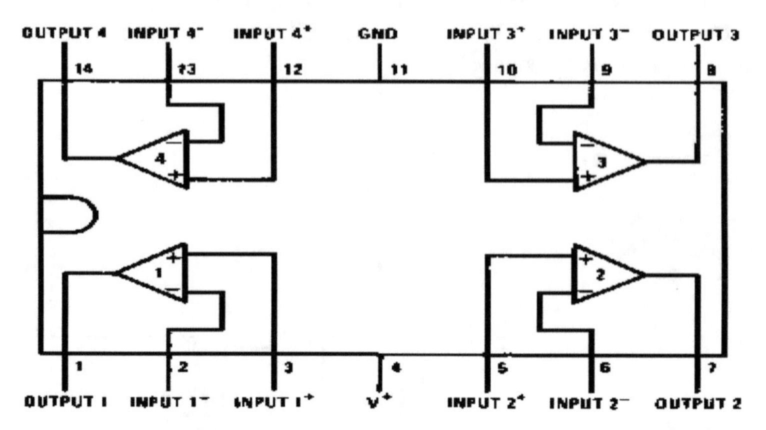

Fig. 12. Internal structure of LM324N.

The TIP36 transistor of Fig. 11 was configured as a switch, 1 kΩ resistor was used as the base resistor R_b and this limits the base current and hence limits voltage drop at the output of the transistor, which switches ON the SSR, the switching inter phase in this case opens and closes based on the signal from the SSR. When the voltage at pin 2 is higher than that at pin 3, the output signal will be zero, then the NOT gate will trigger the LED (yellow) ON. This shows there is no output from the operational amplifier, which shows that there is over voltage and the SSR will cut off the load. Furthermore, the current transformer of Fig. 10 takes in current (30 A maximum) into its input (primary) from the supply (mains), and gives equivalent stepped down voltage (12 V) at the output (secondary), the 12 V AC is rectified using a full wave bridge rectifier and it's then filtered. The output of the filter circuit is fed into the third operational amplifier which also acts as a comparator (current sensing unit) compares voltages between pin 5 and pin 6 of the operational amplifier. The non-inverting terminal pin 6 was also set to a reference voltage of $6V_{DC}$, pin 7 gives output only when pin 5 is higher than pin 6. The 10 kΩ variable resistor is used to set voltage between pin 5 and pin 6, such that pin 5 is higher than pin 6. The output signal from the comparator biases the fourth operational amplifier which is also configured to act as a comparator which equally compares the voltage between pin 10 and pin 9 and gives output at pin 8 when pin 10 is higher than pin 9. The output signal is fed to the second operational amplifier which also compares the voltage from the current sensing unit and the reference voltage of the voltage sensing unit and gives output when the voltages are equal or the voltage from the current sensing unit is less. The output signal from the second operational amplifier flows through the 1 kΩ resistor to the TIP36 transistor, which then switch ON and OFF the solid-state relay, when there is over voltage, over current and/or short circuit, the switching inter phase is opened, hence the load is cut off by the solid-state relay. However, when the fault is cleared, the switching inter phase is closed, thus the solid-state relay triggers the load which makes the system to function normally again. The implemented of the protection system is shown as Fig. 13.

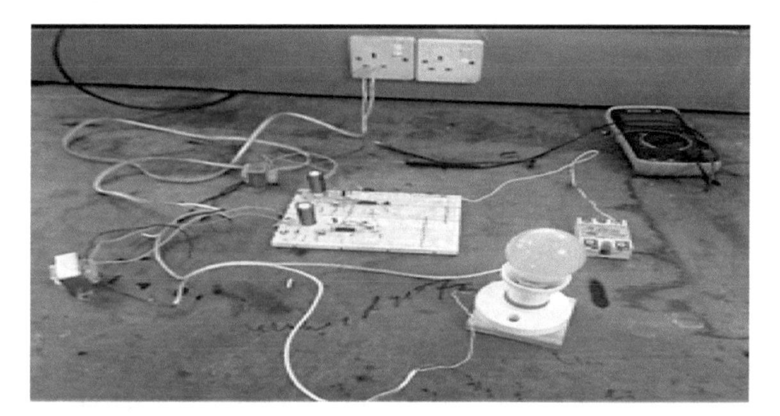

Fig. 13. Diagram showing implementation of the protection system

4 Conclusion

Problem technique plays a fundamental role in guaranteeing the integrity and safe operation of any electrical energy system. The first protection system was based on electromechanical devices employing movable parts and in a later period of development, solid-state based devices with discrete components were introduced. Although both types of devices are still widely used in on-operation protection systems they are currently being replaced by microprocessor-based relays called digital relays. This paper provides the implementation of solid-state relays for enhancement of a perturbed power system protection. With the implementation of the digital technology on the disturbed system, line, transformer, generator of the system was protected to certain integration. The benefit achieved in using numerical relays in this study is accurate tripping and less tolerance display of fault parameters. Result of the study shows that solid-state relays are essential part of the power system and are responsible for the control of any overload voltage or current.

References

1. Mbunwe, M.J., Ogbuefi, U.C., Anyaka, B.O., Ayogu, C.C.: Protection of a disturbed electric network using solid state protection device. In: Lecture Notes in Engineering and Computer Science: Proceedings of The World Congress on Engineering and Computer Science 2018, WCECS 2018, San Francisco, USA, 23–25 October 2018, pp. 249–257 (2018)
2. Rao, S.: EHV-AC, HVDC Transmission and Distribution Engineering Theory, Practices and Solve Problems. Khanna Publishers, Delhi (2006)
3. Power System Protection-Wikipedia. https://en.wikipedia.org/wiki/Power-system_protection
4. Weedy, B.M.: Electric Power Systems, 5th edn. Wiley, Hoboken (2012)

5. Mason, C.R.: The Art and Science of Protective Relaying. General Electric, 20 April 2017
6. Protective Relay-Wikipedia. https://en.wikipedia.org/wiki/Protective_relay
7. Lundqvist, B.: 100 years of relay protection, the Swedish ABB relay History. ABB, 30 May 2017
8. Input interfacing circuits. http://www.electronics-tutorials.ws/io/io_5.htm
9. Rockefeller, G.D.: Fault protection with digital computer. IEEE Trans. Power Appar. Syst. **88**(4), 438–461 (1969)
10. Bo, Z.Q., Jiang, F., Chen, Z., et al.: Transient based protection for power transmission systems. In: IEEE PES Winter Meeting, Singapore (2000)
11. Generator Protection. https://www.electrical4u.com/generator-protection
12. Transformer Protection and Transformer Faults. https://www.electrical4u.com/transformer-protection-and-transformer-fault
13. Busbar Protection and busbar differential protection scheme. https://www.electrical4u.com/busbar-protection
14. Protection of lines or feeder. https://www.electrical4u.com/protection-of-lines-or-feeder
15. Three phase solid state relays – POWERSEM. www.powersem.net
16. Anil Kumar Reddy, K, Sirisha, S.: A ZVS DC-DC converter with specific voltage gain for inverter operation used for 3-phase induction motor operation. Int. J. Adv. Res. Electr. Electron. Instrum. Eng. **2**(8) (2013)
17. Why the use solid state relays. www.crydom.com
18. Shannon, C.E.: A Symbolic Analysis of Relay and Switching Circuits. IEEE Press, Hoboken (1940)
19. Sullivan, K.R.: Understanding relays, autoshop101.com
20. Monseth, I.T., Robinson, P.H.: Relay Systems. McGraw-Hill Book Co., New York (1935)

Optimal Reliability Performance Evaluation of Renewable Energy Power-Driven Power Stations

Uche Chinweoke Ogbuefi(✉) and Muncho Josephine Mbunwe

Department of Electrical Engineering, University of Nigeria, Nsukka, Enugu State, Nigeria
{Uche.ogbuefi,muncho.mbunwe}@unn.edu.ng

Abstract. Electricity generation in Nigeria is based on thermal and hydro generation, and is largely dominated by two sources - non-renewable thermal (natural gas and coal) and renewable water or hydro. Coal and natural gas make up the largest portion of energy production in Nigeria. The fact that a reliable power station is one which would supply the required power within its installed capacity at any time within the specified voltage and frequency limits remains vital to our economic growth, personal well-being. Essential for this evaluation are the station's installed capacity and available generation. As the gap between supply and demand on electric energy amplifies in Nigeria, it becomes necessary to carry out the optimal reliability performance of renewable energy powered turbine generators that control the electric power industry in Nigeria. The work focuses on the past data for performance of some power stations in Nigeria from 2006 to 2017 to determine their reliability and availability of energy within their installed capacities in line with energy global best practices. The combined installed capacity of these power plant is 37% of the twenty-one thermal power plants connected to the national power grid. A historical operational data of these selected plants over a period of twelve years was obtained and evaluated based on power plant performance indices analytical techniques. The relative index of equipment reliability obtained in this study, for Afam I–V, Afam VI, Delta and Egbin are (17.13, 78.57, 34.42, & 70.17) percent accordingly. It also shows that Afam VI and Egbin power stations have good workable preventive maintenance programmes whereas, Afam I–V and Delta power stations were always faced with corrective maintenance.

Keywords: Availability factor · Indices · Optimal reliability · Power stations · Renewable energy · Analytical techniques

1 Introduction

Renewable energy like fossil fuels are hydrocarbons, primarily coal, fuel oil or natural gas, formed from the remains of dead plants and animals. Reliability of electrical power plant is the probability that it will generate electric energy for consumers without interruption and in an acceptable quality in line with designed specifications [1, 2].

S.-I. Ao et al. (Eds.): WCECS 2018, *Transactions on Engineering Technologies*, pp. 139–157, 2020.
https://doi.org/10.1007/978-981-15-6848-0_12

The high rate of electricity demand requires stable and continuous supply of electrical power to consumers. Hence improvement of the operational performance of a nation's electric supply is vital for its economic and social developments. It has been observed that the energy generated by the major hydro-electric power stations in Nigeria does not meet up with the demand [2]. Consumers of electricity both domestic and industrial have been looking forward to improved performance of the available plant/stations from what is presently obtainable. Optimal reliability performance evaluation of a generating system is fundamentally concerned with predicting if the system can meet its load demand adequately for the period of time intended.

Bulk Electric power supply system consist of three functional subunits that could be separately analysed. These three subunits are the power generation, power transmission and power distribution [3, 4]. The work focus mainly on the determination of the reliability of generation system.

The 2014 annual technical report from TCN shows that the total installed capacity of power generation in Nigeria is 11,165.40 MW and that, an average daily capacity of 6,317.70 MW was generated in that year. Out of the average daily load on the national power grid, the hydroelectric.

The most exciting aspect of the problem of power generation in Nigeria is the inability of the power generating companies to maintain the operating power. The selected power plants are Afam I–V, Afam VI, Delta, and Egbin power plants. The years of these power plants covers old generation fossil fuel operated power plants, (had been in operation for over 50 years), middle generation plants (been in operation between 10–48 years) and new generation power (plants commissioned between 1–10 years ago). This represents three generations of thermal power projects in Nigeria. Afam I–V renewable fueled power station falls under the old generation power plant in Nigeria power sector. Afam I–V had an initial installed capacity of 972.8 MW which at present is about 351 MW with twenty power generator units (GT1–GT20). All the generator drivers are simple cycle gas turbines [3, 5]. Ughelli Power Station (formerly called Delta power station) had an initial installed capacity of 912 MW. It also have twenty simple cycle gas turbines generator units (GT1–GT20) initially, the first two generator units were out of service since 2002. The current installed capacity of Ughelli power station is 900 MW.

Egbin Power Station has six renewable energy fired steam turbines generator units (from ST1 to ST6), with total installed capacity of 1320 MW. Each generator set is meant to operate on dual fuel (that is, gas and high pour fuel oil) and have a single reheat and six stages of regenerative feed heating steam generators [6].

Afam VI Power Station belongs to Shell Petroleum Development Company of Nigeria Limited (SPDC), and is being operated by them. The station has three combined cycle gas turbines (labeled GT11–GT13), each rated 150 MW and one 200 MW steam turbine generator (ST1). This gives a total installed capacity in Afam VI power plant as 650 MW. This power station is included in this study to represent the new generation power plants [4, 5].

Power system reliability studies can be conducted for these purposes:

First, long-term reliability evaluations may be performed to assist in long-range system planning;
And secondly, short-term reliability predictions may be sought to assist in day-to-day operating decisions.
Also, valuations of system security where the effects of sudden disturbances are evaluated. Both types of studies may require very different models and mathematical approaches.

2 Concepts of Power Generating Plant Reliability

The ability of a system to consistently perform its intended or required function on demand and without dilapidation or failure is said to be reliable. Power sector is experiencing restructuring era: cheap natural gas, lower cost renewable power sources and increased use of energy efficiency and distributed generation are leading to a transformation. As some of these generators are retiring in recent years and been replaced with new sources of power and energy efficiency, there have been questions about how to sustain the current level of reliability.

In analyzing the power generation indices, the analytical technique of forced outages is adopted in the valuation of the four major selected thermal power generating plants in the Nigerian power sector. The emphasis on examining the performance of thermal power plants is due to the fact that, renewable (fossil) energy power plants constitute 82.7% of the total installed power generation capacity on the national electrical power network. The challenges of extreme electricity shortage over the years has been facing the citizens of Nigeria particularly those in academic. Despite its huge resource and gift of nature in energy and massive investment in providing energy infrastructure, the performance of the power sector has remained poor, in comparison with other developing economies [4, 7, 8]. System components are classified into different sensitive critical levels such that when failures occur, shutdown or just an alarm is triggered. A thermal power generator arrangement consist of several systems, subsystems and auxiliaries that are designed and programed to operate in unison. As a result, component failure rate affects the availability, reliability and capacity utilization of the plant. Reliability assessment on power station are usually tackled from two perspectives; either power plant competence and or power plant security. Power Plant competence is interpreted as having sufficient facilities to generate the required power demand from consumers under static conditions. On the other hand, power plant security hinges on the capability of the plant to absorb both dynamic and transient disturbances prevalent in bulk power supply systems [9, 10].

Reliability of an equipment is the probability that the equipment will sustain operations in accordance with its designed specifications over a given period. Power generation reliability evaluations have been dominated by deterministic and probabilistic methods of modeling [11]. However, deterministic and probabilistic methods of reliability examination/valuation are different but they balance one another [12]. Deterministic reliability assessments are aimed at testing the robustness of delivering stable electric energy in line with standard parameters to consumers under different contingency measures. The deterministic approach requires testing of contingencies by simulating failure of critical components and incorporating sufficient redundancies/backup to prevent those scenarios that could lead to system total collapse [13]. To achieve a standard degree of reliability at the customer level, each of these systems must provide an even higher degree of reliability.

The popular probabilistic indices are: i) LOLP: This describes the probability of the system load exceeding the available generation under the assumption that the peak load of each day lasts all day. It is expressed in units of day/year. ii) LOLE: This describes the expected number of days in a year when loss of load occur. This does not assume that the failure lasts all day. The unit is also in days/year. In practice, the major difference between the two is that in LOLP calculation, daily peak loads are used whereas in LOLE calculation, hourly peak loads are used [13].

The probabilistic modelling method depend on either statistical analysis of data gathered to identify events and the performance of power system components Though probabilistic approach queries the operational data accumulated over the years on the facility to tackle system failures [13, 14]. Probabilistic indices such as, Loss-of-Load Expectation (LOLE), Forced Outage Rates, Loss-of-Load Probabilistic (LOLP), Mean Time between Failure, and Failure Rate, are very popular for evaluating equipment reliability indices. But, the use of probabilistic indices alone is inadequate to determine the reliability of hydrothermal power mix due to sectorial restrains on modelling of hydroelectric and thermal power generation systems. As a result, Equivalent Availability Factor (EAF) was used as the reliability index in computing the operational reliability of the thermal and hydroelectric power plants because, it is impossible to separate the load models for the two different systems that are synchronized onto a common power grid. The research instrument is the Generating Availability Data System (GADS) gathered and compiled in the National Control Centre (NCC) [2, 5, 15]. The electric utility industry initiated GADS in 1982 to expand data collection activities that it began in 1963. Today, NERC's GADS maintains operating histories on more than 5,000 generating units in the North America. GADS is recognized as a valuable source of reliability information for total unit and major equipment groups and is widely used by industry analysts in a variety of applications [12].

3 Methodology

Several methods exist for optimal reliability performance evaluation of renewable energy power-driven Power Stations. In examining the power generation indices, analytical technique of forced outages is adopted in the valuation of the four major chosen thermal power generating plants in the power sector. The number of generator units that were included in the assessment for annual rating of the respective plant are (i) Afam I–V, (ii) Afam VI, (iii) Delta, and (iv) Egbin generator units.

Out of twenty generator units in Afam I–V power station, seven had been scrapped off. Afam VI had four generator units, Egbin had six generator units and Delta out of twenty generator units two are scuffled. Equivalent Availability Factor (EAF) will be used as the reliability index in calculating the functioning reliability of the hydroelectric and thermal power plants.

3.1 Presentation of Data

Parameters like: (i) generator availability; (ii) summaries of the maximum capacities and the average loads of the four chosen power stations. (iii) number of generator trips per year, are the data acquired from GADS-NCC and were used for the valuation of optimal performance indices in the chosen power plants. Generating Availability Data System (GADS) is recognized as a valuable source of reliable information for total unit and major equipment groups, and is widely used by industry analysts in a variety of applications. The summaries of the maximum capacities of the chosen power plants and the average load of each are presented in Table 1. The data collected are presented in Tables 2, 3, 4 and 5 for Afam I–V, Afam VI, Egbin and Delta power plants for eleven years from 2007 to 2017.

3.2 Power System Plants Reliability Indices

Evaluation of the reliability and availability of generator units in the chosen power stations are carried out using the GADS of NCC from 2007 to 2017. Individual generator units' performance of each power plant were obtained. Also, the average performance of all the generator units included in the yearly rating of each power plant gives the plant performance for the specified year. Availability of an equipment is a measure of an operable and committable state of an equipment when it is needed [4]. Each equipment has designed in-built availability (AI) which is defined as:

$$A_r = \frac{MTBF}{MTBF + MTTR} \tag{1}$$

Where: MTBF is Main Time Between failure and MTTR is Main Time to Repair expressed as:

$$MTBF = \frac{Total\ Equipment\ Uptime\ (Days)}{Total\ Number\ of\ Equipment\ Failure} \tag{2}$$

Table 1. Maximum capacity and annual average load summary of the power stations' (in MW)

	Year	2006	2007	2008	2009	2010	2011	2012	2013	2014	2015	2016	2017
Afam I–V	Avr. load (ME)	267.64	152.74	221.40	80.38	228.11	82.12	63.52	21.56	64.84	96.34	57.87	81.08
	Rated capacity (MU)	623.06	623.06	623.06	797.80	931.60	931.60	931.60	516.00	351.00	351.09	351.09	351.09
Afam VI	Avr. load (MU)	NA	NA	NA	NA	NA	56.38	322.82	435.64	486.16	604.70	467.94	554.29
	Rated capacity (MU)	NA	NA	NA	NA	NA	331.50	497.25	650.00	650.00	650.00	650.00	650.00
Delta	Avr. load (MW)	456.67	463.38	393.45	492.49	338.80	211.67	255.33	342.95	246.78	246.23	246.78	409.10
	Rated capacity (MU)	912.00	912.00	912.00	882.00	882.00	882.00	882.00	900.00	900.00	900.00	900.00	900.00
Egbin	Avr. load (MW)	1031.0	1053.48	1147.78	1005.48	735.53	694.97	980.89	819.55	939.11	1022.56	976.77	970.41
	Rated capacity (MU)	1320.00	1320.00	1320.00	1320.00	1320.00	1320.00	1320.00	1320.00	132000.00	1320.00	1320.00	1320.00

Table 2. Afam I–V power station generator uptime (in days)

Unit capacity		Afam-II (44 × 23.9 MW)				Afam-III (2× 27.5 MW)			Afam-IV (54 × 75 MW)				Afam-V (2 × 138 MW)			Total run days				
Unit tag		GT5	GT6	GT7	GT8	GT9	GT10	GT14	GT15	GT16	GT17	GT18	GT19	GT20	Afam-II	Afam-III	Afam -IV	Afam-V	Total days	P/S Aval.
Year	2006	5	150	226	0	0	0	0	0	0	144	0	358	351	376	0	144	709	1229	246
	2007	61	0	0	0	0	0	0	0	0	3	112	348	298	61	0	115	646	822	274
	2008	9	0	0	0	0	0	0	0	0	271	0	46	53	9	0	271	99	379	95
	2009	3	0	0	0	0	0	0	0	0	182	0	0	0	3	0	182	0	185	93
	2010	0	0	0	0	0	0	0	0	0	37	0	0	0	0	0	37	0	37	37
	2011	0	0	0	0	0	0	0	0	0	23	286	0	0	0	0	309	0	309	155
	2012	0	0	0	0	0	0	0	0	0	200	336	0	0	0	0	536	0	536	268
	2013	0	0	0	0	0	0	0	0	0	0	267	0	0	0	0	267	0	267	267
	2014	0	0	0	0	0	0	0	0	0	88	316	0	0	0	0	404	0	404	202
	2015	89	21	0	0	0	0	0	0	0	314	0	161	309	110	0	314	470	1190	238
	2016	0	112	151	0	0	49	0	0	0	284	168	339	313	263	49	452	652	1616	231
	2017	0	266	0	0	0	0	0	0	0	326	0	16	276	7	0	326	292	984	246

Table 3. Afam VI power station generator uptime (in days) with unit capacity of 200 MW

Unit capacity		3 × 150 MW			200 MW	P/S uptime	Total days (150 MW)	Tatal days (200 MW)
Unit tag		GTIl	G-T12	GT13	STI			
Year	2010	310	342	341	N/A	331	093	0
	2011	336	306	298	198	235	940	198
	2012	336	351	360	331	345	1047	331
	2013	334	282	342	286	311	958	286
	2014	335	358	355	317	341	1048	317
	2015	338	385	350	303	307	976	374
	2010	340	333	353	316	337	1049	303
	2017	341	380	351	291	303	994	289

$$MTTR = \frac{Total\ Equipment\ Downtime\ (Days)}{Total\ Number\ of\ Equipment\ Failure} \tag{3}$$

Unavailability or downtime complements availability or uptime. Also, total time is equal to uptime plus downtime. Total Time (1 year) = Uptime + Downtime = Uptime + Downtime (Unplanned + Planned)

$$\text{Uptime} = \text{Total Time} - \text{Downtime (Unplanned + Planned)} \tag{4}$$

Note; Total time applied in this paper is either 365 days (or 366 days if it's a leap year). The annual trip data for the considered power stations are represented in Tables 6, 7, 8 and 9 accordingly.

To illustrate the evaluation of generator unit of Afam I–V's MTBF and MTTR using Eqs. (2 and 3): From Table 2, GT6 and GT7 operated for 226 days and 0 day in 2006 respectively. From Table 6, GT6 had 8 trips in 2006 whereas, GT7 had 0 trip. Therefore, the MTBF and the MTTR of GT6 and GT7 in 2006 are calculated as follows:

$$MTBF\ for\ GT6 = \frac{226\ (Days)}{8} = 28\,days,\ \text{and}$$

$$MTTR\ for\ GT6 = \frac{(365 - 226)\ (Days)}{0} = 0\,days$$

$$MTBF\ for\ GT7 = \frac{0\ (Days)}{0} = 0\ days,\ \text{and}$$

$$MTTR\ for\ GT7 = \frac{365 - 0\ (Days)}{0} = 0\ days$$

$$MTBF\ for\ GT17 = \frac{144\ (Days)}{3} = 48\,days\ \text{and}$$

$$MTTR\ for\ GT6 = \frac{(365 - 144)\ (Days)}{3} = 74\,days\ \text{etc.}$$

Table 4. Delta power station generator uptime (in days)

Unit capacity		Delta-II (6 × 25 MW)							Delta-III (6 × 25 MW)						Delta-IV (6 × 1OO MW)					Total days (25 MW)	Total days (100 MW)	P/S Aval.
Unit tag		GT3	GT4	GT5	GT6	GT7	GT8	GT9	GT10	GT11	GT12	GT13	GT14	GT15	GT16	GT17	GT18	GT19	GT20			
Year	2007	0	0	0	358	311	344	353	348	355	358	351	282	90	0	253	358	0	316	3060	1017	314
	2008	0	0	0	102	79	121	313	213	343	291	0	324	163	0	0	25	0	226	1786	414	200
	2009	0	0	0	102	63	10	215	187	262	236	0	236	0	0	0	295	0	333	1421	628	205
	2010	0	0	0	251	307	78	326	125	324	349	57	269	51	276	49	270	199	148	2086	993	205
	2011	0	0	0	63	42	65	206	103	320	209	135	197	136	302	135	114	237	294	1340	1218	171
	2012	0	349	0	0	0	318	366	0	309	295	0	0	0	0	0	76	300	296	1637	672	289
	2013	0	175	0	0	0	331	0	0	0	246	0	0	0	183	126	0	73	349	752	731	212
	2014	0	365	0	0	365	346	0	323	275	328	92	0	0	336	363	0	0	351	2094	1050	314
	2015	0	245	0	0	338	339	0	271	61	47	42	0	0	225	346	0	0	106	2385	1174	209
	2016	0	243	351	21	365	358	0	9	9	34	32	0	0	265	365	0	0	116	2412	1221	319
	2017	0	231	0	0	365	336	0	9	11	35	32	0	0	184	136	0	0	243	2029	1059	157

Table 5. Generator uptime (in days) for Egbin power station

Unit capacity		6 × 220 MW							
Unit tag		ST1	ST2	ST3	ST4	STS	ST6	Total	P/S Avail
Year	2007	277	351	23	337	63	30	1356	271
	2003	316	246	94	276	331	0	1263	253
	2009	312	354	302	331	310	0	1609	322
	2010	24	351	346	358	338	0	1417	283
	2012	360	356	313	327	320	0	1676	335
	2013	340	363	340	328	355	0	1726	345
	2014	307	339	313	343	299	0	1601	320
	2015	322	344	347	314	279	0	1606	321
	2016	337	335	347	357	343	352	2067	346
	2017	323	335	334	347	355	357	2078	346

Table 6. Afam I–V power station generator unit annual trips

Unit capacity	Afam-II (4 × 23.9 MW)				2 × 27.5 MW			Afam-IV (5 × 75 MW)				2 × 138 MW	
Unit tag	GT5	GT6	GT7	GT8	GT9	GT10	GT14	GT15	GT16	GT17	GT18	GT19	GT20
2006	4	8	NA	NA	NA	NA	NA	NA	NA	3	NA	1	3
2007	5	1	NA	NA	NA	NA	NA	NA	NA	3	3	2	2
2008	2	NA	NA	NA	NA	NA	NA	NA	NA	7	NA	2	2
2009	1	NA	NA	NA	NA	NA	NA	NA	NA	19	NA	NA	NA
2010	NA	NA	NA	NA	NA	NA	NA	NA	NA	NA	4	NA	NA
2011	NA	NA	NA	NA	NA	NA	NA	NA	NA	7	18	NA	NA
2012	NA	NA	NA	NA	NA	NA	NA	NA	NA	23	15	NA	NA
2013	NA	NA	NA	NA	NA	NA	NA	NA	NA	NA	15	NA	NA
2014	NA	NA	NA	NA	NA	NA	NA	NA	NA	7	12	NA	NA
2015	NA	1	NA	NA	NA	NA	NA	NA	NA	13	12	NA	NA
2016	NA	NA	NA	NA	NA	NA	NA	NA	NA	1	14	NA	2

Table 7. Afam VI power station generator units annual trips.

Unit capacity		3 × 150 MW			200 MW
Unit tag		GT11	GT12	GT13	ST1
Year	2010	16	8	11	NA
	2011	7	8	10	15
	2012	2	3	3	4
	2013	2	4	6	10
	2014	5	5	5	6
	2015	9	6	5	7
	2016	4	8	7	3
	2017	7	6	8	9

Also, the MTBF and MTTR of other generator units in all the four stations were calculated for the eleven years period of this work. The calculated MTBFs for Delta and Egbin are used to plot the charts as shown in result section.

3.3 Plant Energy Availability Factor (PEAF)

Station Plant Equivalent/Energy Availability Factor over one year period 'f' is the ratio of energy 'H' that the available capacity (h) could have produce during one year to the energy 'G' that the maximum capacity (g) could have produced in that same year:

$$EAF : f = \frac{H}{G}\text{(expressed in percentage of the energy G)} \tag{5}$$

The energies G and H are expressed mathematically as:

$$H = \Sigma h.dt \text{ or } H = \Sigma h.th, \quad \text{and} \quad G = \Sigma g..dt, \quad or \quad G = \Sigma g.tg \tag{6}$$

Where: th = duration of available capacity h and tg = duration of maximum capacity g as in (5),

$$PEAF = \frac{Plant\ Average\ Load\ (PAL)\ MW\ in\ a\ given\ Year}{Plant\ Maximum\ Capacity\ of\ the\ Plant\ (PMC)\ MW\ in\ that\ given\ Year} \tag{7}$$

Calculation of PEAF is carried out using Eq. (7). From Table 1 Afam I–V PEAF for 2012 is calculated thus

$$PEAF\ for\ Afam\ I - V = \frac{95.32\,\text{MW}}{351.00\,\text{MW}} = 0.272,$$

$$PEAF\ for\ Delta = \frac{246.23\,\text{MW}}{900.00\,\text{MW}} = 0.274,$$

$$PEAF\ for\ Afam\ VI = \frac{603.70\,\text{MW}}{650.00\,\text{MW}} = 0.929, \text{ and}$$

$$PEAF\ for\ Egbin = \frac{1022.56\,\text{MW}}{1320.00\,\text{MW}} = 0.775$$

Table 8. Delta power station generator units yearly trips

Unit capacity		Delta-II (6 × 25 MW)							Delta-III (6 × 25 MW)						Delta-IV (6 × 100 MW)				
Unit tag		GT3	GT4	GT5	GT6	GT7	GT8	GT9	GT10	GT11	GT12	GT13	GT14	GT15	GT16	GT17	GT18	GT19	GT20
Year	2007	NA	NA	NA	1	3	3	1	2	1	1	2	2	1	NA	2	1	NA	3
	2008	NA	NA	NA	1	NA	3	3	5	2	1	NA	2	4	NA	NA	2	NA	3
	2009	NA	NA	NA	6	4	1	6	9	7	6	NA	4	NA	NA	NA	6	NA	3
	2010	NA	NA	NA	19	14	9	7	11	11	6	5	11	6	10	6	18	12	17
	2011	NA	NA	NA	3	8	9	7	8	4	11	7	8	6	6	7	8	10	12
	2012	NA	6	NA	NA	NA	5	1	NA	2	9	NA	NA	NA	NA	NA	2	19	12
	2013	NA	7	NA	NA	NA	15	NA	NA	NA	15	NA	NA	NA	5	11	NA	3	9
	2014	NA	1	NA	NA	1	1	NA	3	1	5	1	NA	NA	3	2	NA	NA	3
	2015	NA	8	NA	NA	3	2	3	8	NA	6	3	NA	NA	5	8	3	NA	9
	2016	NA	2	NA	NA	4	6	NA	2	4	6	2	5	1	3	7	NA	8	7
	2017	0	2	NA	NA	3	4	6	NA	2	4	6	3	1	2	4	NA	5	3

Table 9. Egbin power station generator units annual trips

Unit capacity		6 × 220 MW					
Unit tag		ST1	ST2	ST3	ST4	ST5	ST6
Year	2010	3	9	6	4	6	NA
	2011	4	9	16	12	12	NA
	2012	6	3	13	10	3	NA
	2013	10	6	9	6	9	NA
	2014	5	4	7	6	10	NA
	2015	6	11	10	5	4	NA
	2016	5	7	6	4	7	NA
	2017	7	4	7	6	4	NA

Also, the annual data in Table 1 and 5 have been used to calculate the yearly PEAF for the four considered power plants. The results of the annual PEAF of four chosen plants under review are presented in Fig. 1.

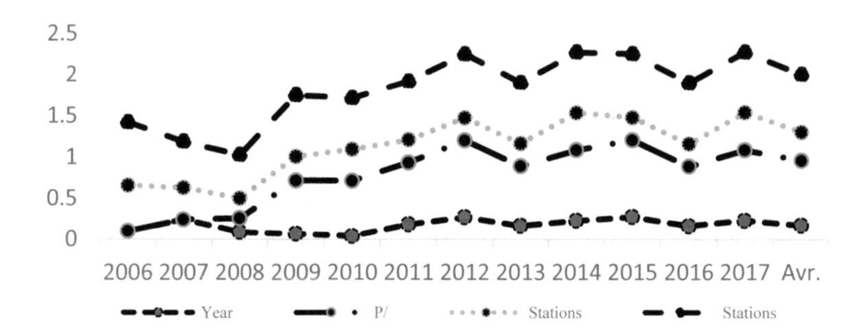

Fig. 1. Energy/equivalent availability factors for the four power stations

Plant Energy Availability Factor (PEAF) takes into account the health of the generators and de-rated generator units of the plant and hence, it models both the partial and full outages of the generators in the plant.

Thus, this index gives the true measure of the probability of the power station performing its intended function. Energy Available Factor (EAF) illustrates the reliability of the plant in general, taking into account all complete and partial outages [2, 4, 9].

Generator Equivalent Availability Factor (GEAF) is expressed as:

$$GEAF = \frac{Generator\ Average\ Load\ (GAL)\ MW\ in\ a\ given\ Year}{Generator\ Maximum\ Capacity\ \ (GMC)\ MW\ in\ that\ given\ Year} \tag{8}$$

3.4 Model for Calculation of Generator Average Load (GAL) for the Given Year

From the generators operational Uptime Table for the given plant, separate and add up the total Uptime for the generators with similar installed capacities within the year as presented at the extreme right of the uptime tables in Tables 2. For Afam I–V plants, summation of the generator units with the same nameplate and capacities that contributed to the annual maximum rating of the plant were carried out. Equation (9) is the developed model for determining the Generator Average Load (GAL) from the weighted Plant Average Load (PAL) as presented in Table 3.

$$GAL = \frac{PAL\ (MW)\ \times\ TCSU\ (MW)}{PMC\ (MW)} \times \frac{Uptime\ of\ the\ Unit(Days)}{Total\ Uptime\ of\ Similar\ Units\ (Days)} \quad (9)$$

Where: TCSU = Total Capacity of Similar Units (MW) operated in the year & PMC = Plant

Rated (maximum) Capacity of the year.

Calculation of generators annual/yearly average load are as illustrated below.

Taking 2007 as reference, from Table 2, the Uptime for Afam I–V GT5, GT18, GT19 and GT20 are (61, 112, 348 and 298) days respectively. One GT5 operated in the year, two similar sizes of GT18 and two similar sizes of GT20. Total Uptime of similar Units (in days) are shown at the extreme right columns of generators in Table 2. The weighted average Load of Afam I–V in the year 2007 & 2008 are 228.11 MW & plant rating of 931.6 MW, 82.12 MW & plant rate of 931.6 MW as shown in Table.

Using Eq. (9), the Average Loads carried by each generator unit are calculated thus: For 2007, we have GT5, GT17, GT18, GT19 and GT20 respectively.

$$GAL\ for\ GT5 = \frac{228.11\,\text{MW} \times 47.8\,\text{MW}}{931.60\,\text{MW}} \times \frac{61\ (Days)}{61\ (Days)} = 11.70\,\text{MW}$$

$$GAL\ for\ GT18 = \frac{228.11\,\text{MW} \times 150\,\text{MW}}{931.60\,\text{MW}} \times \frac{112\ (Days)}{115\ (Days)} = 35.77\,\text{MW}$$

$$GAL\ for\ GT19 = \frac{228.11\,\text{MW} \times 276\,\text{MW}}{931.60\,\text{MW}} \times \frac{348\ (Days)}{646\ (Days)} = 36.41\,\text{MW}$$

For 2008 we have GT5, GT17, GT19, & GT20

$$GAL\ for\ GT19 = \frac{82.12\,\text{MW} \times 276\,\text{MW}}{931.60\,\text{MW}} \times \frac{46\ (Days)}{99\ (Days)} = 11.30\,\text{MW}$$

$$GAL\ for\ GT20 = \frac{82.12\text{MW} \times 276\,\text{MW}}{931.60\,\text{MW}} \times \frac{53\ (Days)}{99\ (Days)} = 11.30\,\text{MW etc.}$$

Likewise, the Generator Average Load (GAL) for all the generator units in the four power stations are calculated and the graphs are as presented Sect. 4.

Subsequently, computing GAL for Afam I–V's GT6, GT10, GT17 and GT20 in the above examples, Values of the Generator Average Load (GAL) obtained from Eq. (9) are substitute into Eq. (8) to get the value of equivalent availability factor of each generator unit. For 2015 & 2016:

$$GEAF_{GT7} = \frac{10.69\,\text{MW}}{23.9\,\text{MW}} = 0.45; \quad GEAF_{GT16} = \frac{29.50\,\text{MW}}{75\,\text{MW}} = 0.39;$$
$$GEAF_{GT17} = (\frac{45.40}{75})\,\text{MW} = 0.61;$$
$$GEAF_{GT20} = (\frac{17.09}{138})\,\text{MW} = 0.05;$$
$$GEAF_{GT5} = (\frac{5.96}{23.9})\,\text{MW} = 0.25; \qquad GEAF_{GT7} = (\frac{6.20}{23.9})\,\text{MW} = 0.26;$$
$$GEAF_{GT17} = (\frac{19.93}{75})\,\text{MW} = 0.27;\ \text{etc.}$$

These calculated (GAL) (in tables-skipped) are used to generate the graphs as presented Sect. 4.

4 Results and Analysis

After careful Performance Evaluation study for the considered stations, the results are as presented in Figs. 2, 3, 4, 5 and 6. Suffice it to say that the main reliability indices studied are operational uptime/availability, energy availability of the power stations, the main time between failures, and the main time to repair.

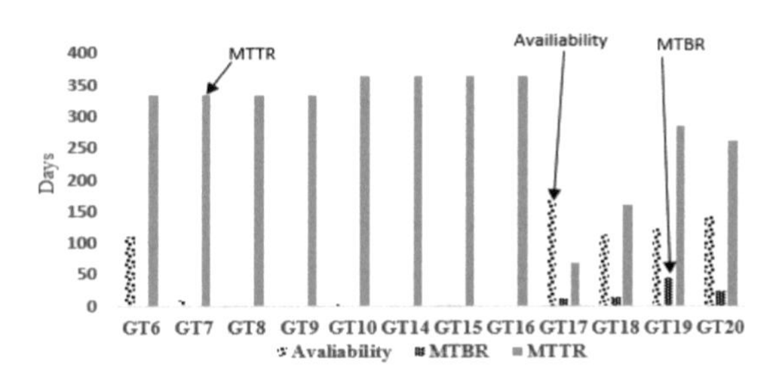

Fig. 2. Average performance on reliability indices by generators in Afam I–V power plant (for 11 years)

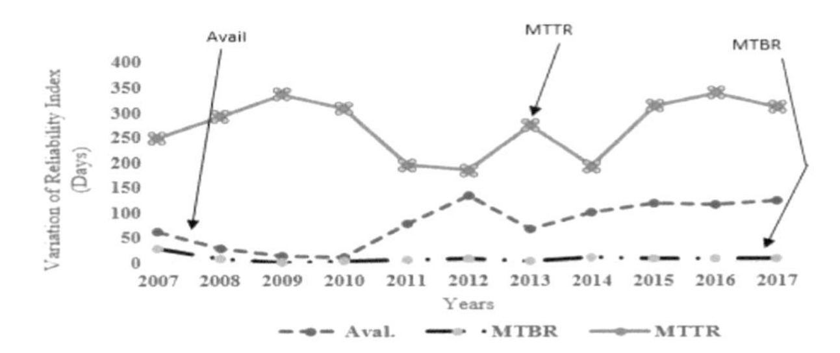

Fig. 3. Reliability indices variation of Afam I–V power plant with year.

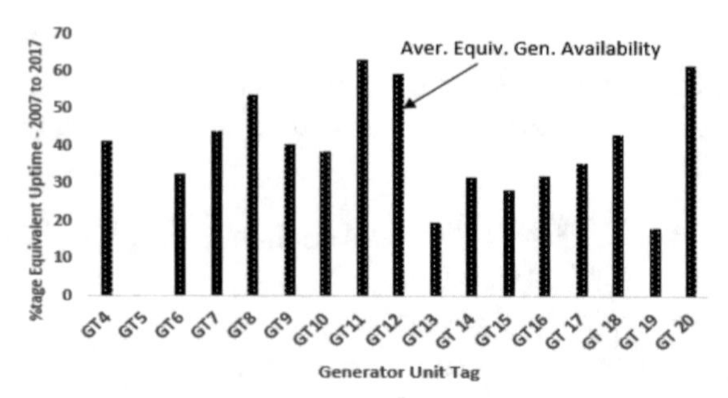

Fig. 4. Average equivalent availability of generator units in delta power station (11 years)

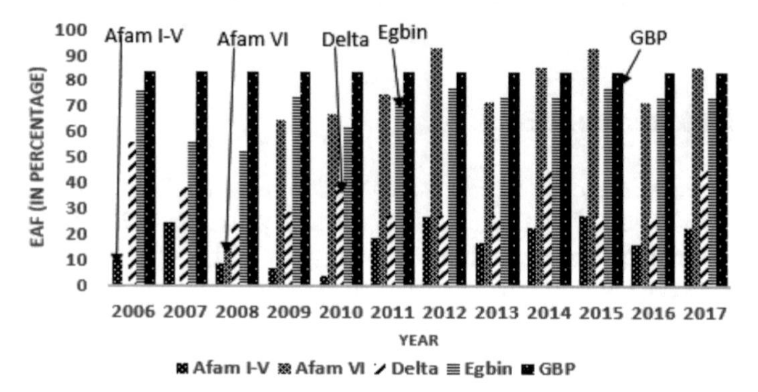

Fig. 5. Plant equivalent average factor (PEAF) variation for the four considered power stations with year

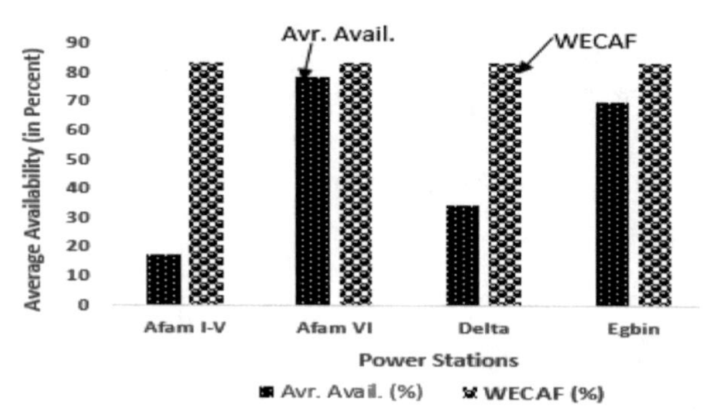

Fig. 6. Station average availability performance vs world energy council availability factor (WECAF)

Calculated Reliability indices in Afam I–V Power Station Presented in Tables 2, 10 and 12 (Tables 10 and 12 skipped), is shown in Fig. 2. Furthermore, with the Afam IV calculated reliability indices shown in Tables 2, 10 and 12, the average yearly performance of power station is generated and its as shown in Fig. 2.

Figure 2 shows that much time was devoted to generator repairs hence unit's unavailability in the period of this study, in contrast to running the units.

From Fig. 1, it could also be seen that GT17 had the lowest average turnaround maintenance period (MTTR) of 196days with average MTBF of 14days. Although GT17 came out as the unit with the highest availability, followed by GT20 and GT19, the characteristics of its reliability indices is inadequate.

Figure 3 shows that within the period (12 years) of the study, more time was spent in breakdown maintenance on generator units. The graphs also shows that availability of the plant have inverse relationship with MTTR. It also confirmed that Afam I–V was affected with constant turbine failures with prolong downtime of generator units and that there are not enough evidence of preventive maintenance activities in the power station. Thus this affected plant operation and energy availability.

The calculated equivalent availability data in Table 18 (skipped) is used to generate the energy availability graph for the generator units in Delta power station.as shown in Fig. 4.

From the pictorial graph, Generator unit 8 has the best performing unit during the period under the study with average reliability of 63%. This is followed by GT20 and GT12 with average percentage values of 61.8 & 59.7. GT5 and GT19 are the least reliable unit during the period under the study. [Note: Due to limitation of space, some stations detailed analysis were skipped. Contact the Authors iff.]

4.1 Reliability Analysis of the Four Studied Power Stations

From the evaluated data and information, the graph in Fig. 5 is produced. It shows the percentage equivalent availability of the four power stations for the eleven years period of the study (from 2007 to 2017). Having implemented the Equivalent Availability as the relative index of quality reliability in this study, the reliability of the four thermal power plants varies from (4.18 to 27.44)% for Afam I–V, (17.10 to 93.03)% Afam VI, (24 to 55.84)% Delta and (52.65 to 77.49)% Egbin power plants respectively.

Figure 5 represents the Plant Energy Availability Factor (PEAF) for eleven years period of study for Afam I–V, Afam VI, Egbin and Delta power stations. Their average optimal reliability performance factors are (17.13, 78.57, 34.42, & 70.17) percent accordingly. The World Energy Council Availability Factor (WECAF) accepted for used as a benchmark value for good performance in Nigeria is 83.50%. This value is juxtaposed into Fig. 5 to compare the optimal performance of the four power plants. The optimal reliability performance factor is as shown in Fig. 6. It was observed that each of the four power plants needs some improvement on their daily availability [16]. We noticed squat falls from the operational equivalent availability of 66.37% by Afam I–VI, 49.08% by Delta power station, 4.93% by Afam VI and 16.67% by Egbin power station respectively. With these it is obvious that certain basic functions are not fully implemented; i) shortage and obsolete machines/equipment, ii) lack of proficient and trained workers conversant with fault location and troubleshooting through the Human-Machine-Interface of the turbine packages. iii) Low operational availability of power plants caused by lack of strategic planning of maintenance activities and poor maintenance practices.

5 Conclusion

Average reliability of Afam I–V, AfamVI, Delta and Egbin are (17.13, 78.57, 34.42, & 70.17) percent respectively. Figure 6 depicts the histogram of the scenario. These values are lower than the WEC factor recommended average energy availability of renewable energy turbine generators. From Fig. 6, the performance of Afam VI and Egbin Power stations could be rated as reasonable while the performance of Afam I–V and Delta power stations are so low. Utility companies and operators of power stations have duty to manage electrical assets in a manner that would guarantee uninterrupted electricity supply and the maintenance of the as built technical reliability of the equipment throughout its life span. Thus, the optimal reliability performance valuation of renewable energy power stations will be highly improved, hence availability of power for enhance economy development.

We have the following recommendations.

1. Competent and qualified workforce that will support the operation and handling of power plant gadgets in the power industry promptly should be need engaged.
2. All aging workforce in the electric power industry coupled with the local content policy of the federation should be properly addressed.
3. Efficiency of aged equipment reliability should be improve by engaging the original equipment manufacturers' in line with terms and conditions of the purchase, care, and improvement agreements.
4. Root cause failure analysis (RCFA) should be carried out for all major equipment failures to dissect underlying causes of defects thereby helping to implement corrective actions to avoid reoccurrence. RCFA functions are to determine the cause of a problem and implement corrective and curative actions efficiently in cost effective manner, to rectify, identified problem and to provide data that can be used for rectifying similar problems in the future.

References

1. Fossil | Department of Energy. https://www.energy.gov/science-innovation/energy-sources/fossil
2. Ogbuefi, U.C., Mbunwe, M.J., Ogbogu, O.N.: Assessment of operational reliability of some fossil energy driven power stations. In: Proceedings of the World Congress on Engineering and Computer Science 2018. Lecture Notes in Engineering and Computer Science, San Francisco, USA, 23–25 October 2018, pp. 263–270 (2018)
3. Gupta, S.A., Tewari, C.P.C.: simulation modelling and analysis of a complex system of a thermal power plant. J. India Eng. Manag. **2**(2), 387–406 (2009)
4. Etebu, O.M.O., Kamalu, U.A., Agara, M.P.: Reliability assessment of selected fossil fuel operated power stations in Nigeria. Int. J. Emerg. Trends Sci. Technol. v3i11.01, 4716–4735 (2016)
5. TCN (Transmission Company of Nigeria), Grid Operations, Annual Technical Report (National Control Centre, Osogbo) (2014)
6. Nagrath, I.J., Kothair, D.P: Modern Power System Analysis, pp. 270–277. Tata McGraw –Hill Publishing Company Ltd., New York (1991)

7. Kola, S., David, M.O.: Privatization and trend of aggregate consumption of electricity in Nigeria: an empirical analysis. Afr. J. Account. Econ. Financ. Bank. Res. **3**(3), 18–27 (2008)
8. Momoh, J.A.: Electric Power System Application of Optimization, pp. 2–12. Marcel Dekker Inc, New York (2001). University, Washington D.C.
9. Akhavein, A., Fotuhi-Firuzabad, M., Billinton, R., Farokhzad, D.: Adequacy equivalent development of composite generation and transmission systems using network screening. IET Gener. Transm. Distrib. **5**(11), 1141–1148 (2011)
10. Gupta, G.R.: Power System Analysis and Design, 6th edn, pp. 614–628. S. Chand & Company Ltd., New Delhi (2012)
11. NERC (North America Electric Reliability Corporation), Bulk Power System Planning for Reliability. Reliability Assessment Guidebook v 3.1, Atlanta, 7–11 2012. http://www.nerc.com
12. Owe, G.I., Inyama, K.: Adequacy analysis and security reliability evaluation of bulk power system. IOSR J. Comput. Eng. **11**, 26–35 (2013)
13. Okonkwo, R.C.: Thermal power stations' reliability evaluation in a hydrothermal system. Niger. J. Technol. **18**(1), 8. ENS (European Nuclear Society), Availability Factor (2015). www.eurosnuclear.org/info/encyclopedia/availabilityfactor.htm
14. Wood, A.J., Wallenberg, B.: Power System Generation, Operation and Control, 2nd edn., pp. 283–288. Wiley, Hoboken (1996)
15. Adamu, M.Z., et al.: Reliability evaluation of Kainji hydro-electric power station in Nigeria. J. Energy Technol. Policy **2**(2), 15–32 (2012)
16. WEC (World Energy Council): Availability and Unavailability Factors of Thermal Generating Plants: Definitions and methods of Calculation (2001). www2.osnerg.gob.pe/procreg/tarifasbarra/ProcNov03-Abr04/pre-est-tec/vol2/7.pdf

Boiling and Thermohydraulics Within Pressure Vessels

Ian Bradley[1], Frank Otremba[1(✉)], Giordano Emrys Scarponi[2], and Romero-Navarrete José-Antonio[1]

[1] Federal Institute for Materials Research and Testing (BAM), Unter den Eichen 44-46, 12203 Berlin, Germany
{Ian.Bradley,Frank.Otremba,Jose-Antonio.Romero-Navarrete}@bam.de
[2] The University of Bologna, Via Zamboni, 33, 40126 Bologna, BO, Italy
giordano.scarponi@unibo.it

Abstract. Exposure of pressure vessels to fire can result in catastrophic explosion and escalation of accidents. The safe transportation of cargo in pressure vessels therefore requires knowledge of what will happen to the cargo in the event of a vehicle derailment or rollover resulting in fire exposure. The chapter presents an overview of selected testing and modelling work undertaken to understand the thermohydraulic processes within a vessel that drive pressurization during fire. A series of experiments highlighting the importance of adequate design and selection of protection systems are summarized. It is concluded that pressure relief alone is typically insufficient to prevent vessel rupture, but the combination of relief and thermal coatings can be effective.

Keywords: BLEVE · Explosion · LPG · PIV · Pressure vessel · Thermohydraulics

1 Introduction

Thermohydraulics is the combined study of fluid mechanics, heat transfer and phase change in fluids, or to put it simply, the study of phenomena when a fluid is exposed to a heat source. One of the main concerns with hazardous fluids and stored within a pressure vessel is the rate of pressure rise, potentially leading to catastrophic failure of the vessel. Fire is a potential source of heat during transportation of dangerous goods due to the risk of derailment or roll-over.

Examples of two such catastrophic events are listed in Table 1 [1, 2]. In both of these cases, one involving a road tanker and the other a railway tanker, containers were subject to direct fire and exploded.

The study of the phenomena involved in such circumstances is therefore of interest to institutions ranging from academia to state bodies and standards setting organization responsible for delivering rules and criteria to ensure the safe design and operation of equipment involved in the transport and storage of hazardous fluids.

In Germany, the Federal Institute for Materials Testing and Research (BAM) has this responsibility. This institute dates back to 1871 and was founded as the Royal

S.-I. Ao et al. (Eds.): WCECS 2018, *Transactions on Engineering Technologies*, pp. 158–172, 2020.
https://doi.org/10.1007/978-981-15-6848-0_13

Table 1. Two major accidents involving spill and explosion of cargo

Country/Year	Fatalities/Injuries/Evacuated people	Description
Italy/2009	22/27/1000	Derailment of an LPG rail tankcar
Spain/1978	200/400/-	Rupture of an overfilled propylene road tanker

Prussian Mechanical Testing Institute. It now covers areas as diverse as materials, energy environment, infrastructure and analytical sciences [3].

BAM has a test facility dedicated to experimental research of significant hazards, known as the technical safety test site (TTS). Fire test facilities at TTS allow for both destructive and non-destructive testing. Figure 1 illustrates the non-destructive testing facility. It is capable of replicating hydrocarbon "pool" type fires using propane burners and subjecting objects with masses of up to 200 tons to representative heat loads, including those specified by the standard UN 1965 [4].

These two facilities are complemented by a jet-fire facility (see Fig. 2b), recently developed to allow a broader spread of fire exposure conditions to be recreated. Figure 3 shows an approximate distinction between fire types [5], however it should be taken as a guide, given that heat flux alone does not determine the severity of a fire exposure and heat fluxes can vary significantly with fuel type and the degree of confinement.

Destructive tests of vehicles up to 40 tons and non-vehicle objects of up to 20 tons can be undertaken on a nearby facility capable of applying fire via a 12 m × 8 m bed of regularly spaced propane burners (see Fig. 2a).

Fig. 1. Non-destructive test facility at BAM

Fig. 2. (a) Destructive test facility at BAM; (b) Jet fire test facility

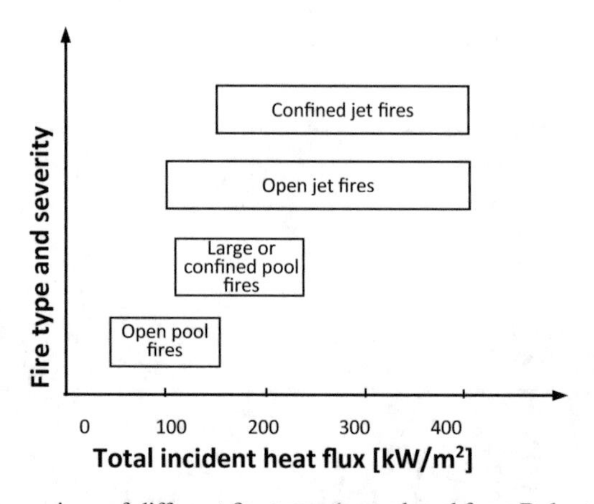

Fig. 3. Comparison of different fire types (reproduced from Roberts et al. [5])

This chapter describes a number of experimental and theoretical methods used to undertake research in the field of thermohydraulic response of pressure vessels to fire. Novel testing equipment and methods of modeling vessel pressurization during fire are briefly discussed. Experimental results concerning the effectiveness of pressure relief valves (PRV) and insulation are given, to further illustrate the practical measures necessary to mitigate the risk of vessel rupture. Advanced modelling of vessels using computational fluid dynamics (CFD) is discussed, together with selected CFD simulation results performed by the University of Bologna, to make recommendations for future research. This paper is based upon a WCECS 2018 Conference paper [30].

2 Study of Vessel Thermohydraulic Behaviour

A novel test facility has been created at BAM, in conjunction with the University of Edinburgh, U.K., and Queen's University, Ontario, with the objective of direct visualization and study of fluid behavior during fire exposure [6, 7]. The purpose of directly studying fluid behavior and wall boiling is to generate the datasets necessary to improve and validate CFD models of vessel response to fire. The facility is referred to as the thermohydraulic monitoring test rig (TMTR). Figure 4 shows the equipment, the main features of which are:

- A one metre diameter test vessel
- A flexible and customizable fire system using propane burners
- A full-bore glass window separating the vessel into a 'test' end and an 'equipment' end
- A pressure compensation to match the pressure across the glass window and prevent breakage during live tests
- A flame shield to protect the window and equipment

a) lateral view of the TMTR equipment

b) front view of the TMTR equipment

Fig. 4. Novel test facility at BAM for studying vessel thermohydraulic response to fire

The TMTR includes extensive instrumentation: 100 thermocouples (throughout the wall, vapor, liquid boundary zones and liquid bulk), pressure transducers, directional flame thermometers, and a particle imaging velocimetry (PIV) system. PIV is an experimental technique designed to capture flow patterns across a wide field of measurement

[8]. A laser, lenses and a mirror were used to create a 2D light sheet perpendicular to the vessel axis. Seeding particles were added to the fluid and the scattered light was recorded by cameras. Dedicated software was then used to analyze the images, giving velocity vector plots for the fluid within the vessel as a function of time. Figure 5a shows a typical PIV image taken at mid height. The vessel wall can be seen as an illuminated arc on the left side. Figure 5b shows a typical vector plot within the first 120 s of a fire test, clearly showing the velocity boundary layer and eddy formation. PIV analysis is also capable of giving streamlines and turbulence statistics. Combining PIV results with the temperature and pressure reading allows for the study of thermal and velocity boundary layers, thermal stratification and other parameters important for vessel pressurization models.

3 Modelling of Vessel Pressurisation

Predicting the catastrophic failure of pressure vessels in fire requires analysis of the loss of strength in the vessel wall and the pressure rise. The former can be modelled with reasonable confidence using either finite element analysis (FEA) or with simpler techniques such as the von-Mises stress. Stress-rupture should also be considered as a mode of failure at higher temperatures [9, 10].

Predicting the rate of pressure rise is significantly more complicated. Over the years, increasingly complex models have been proposed by several researchers. These aim to reproduce the phenomena occurring inside the vessel under fire exposure to predict the pressurization rate, temperature distributions, and ultimately time to failure. Early approaches were based on strong simplifying assumptions [11–20] whereby the tank is divided in one or more zones (or nodes), for which integral mass and energy balances are solved.

Most zone models aim to predict the temperature of the liquid surface and typically calculate the vessel pressure from the vapor saturation pressure. One-zone models assume isothermal contents, whereas two- (or more) zone models include the effects of thermal stratification by dividing the liquid into a hot boundary zone and a sub-cooled core (see Fig. 6). More advanced models may include a greater number of zones to handle phenomena such as vapor super-heating.

A significant deficiency of one-zone models is their inability to account for the rapid rise in pressure due to thermal stratification before PRV action (if one is present) mixes the liquid and it becomes increasingly homogenous [21]. Multi-zone models represent a significant improvement over single-zone models, however they remain dependent on empirical calibration data, particularly regarding the distribution of thermal energy between zones, and struggle to model partial heating of the vessel and other three-dimensional effects.

This limits their range of applicability and the possibility of using such models far from the experimental conditions used for their development. Furthermore, they often neglect key phenomena such as boiling.

More recently, several authors proposed CFD as a promising tool to improve modelling capabilities [22–25]. In this sense, one of the most advanced approaches is the model presented by Scarponi et al. [26]. They used Ansys Fluent to recreate a 2D cross-section of numerous vessel sizes and fill levels. The volume of fluid (VOF) was selected

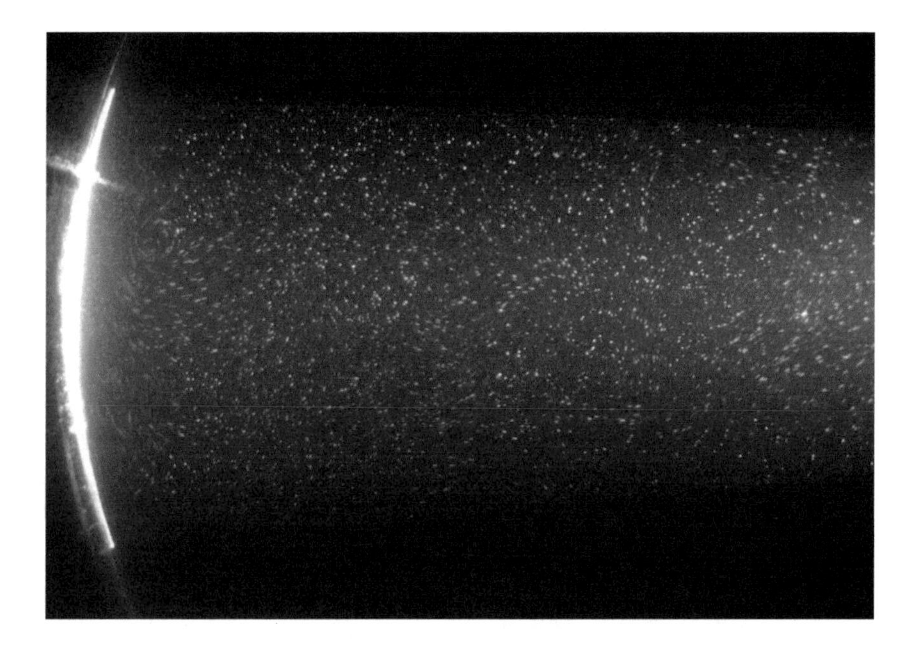

a) typical PIV image for 62%-fill water test, incident heat flux 67 kW/m2

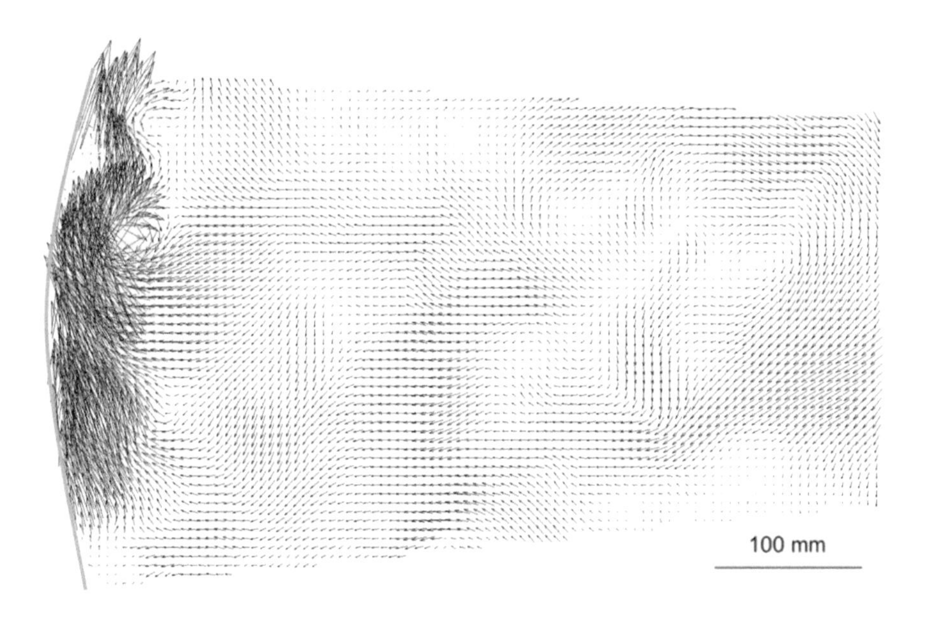

b) vector plot showing velocity distribution and eddy occurrence

Fig. 5. Typical PIV image and vector plot

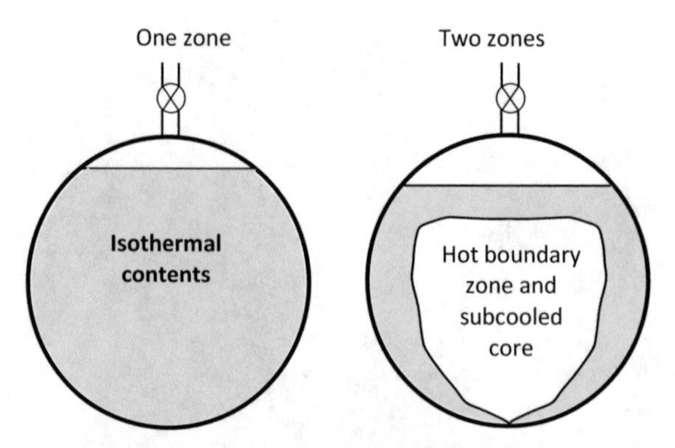

Fig. 6. Schematic diagram of zone-modelling approaches

as the multiphase model, coupled with the Hertz-Knudsen phase change mechanism. Turbulence was considered using the k-ω turbulence model.

The modelling approach was validated against data from an extended set of fire tests simulating full engulfing pool fire scenarios affecting liquefied petroleum gas (LPG) tanks. Simulations results proved to be in good agreement with experimental measurements in terms of pressure and temperatures, demonstrating the robustness of the modelling setup.

Analysis of their results provides an examination of the mechanisms that drive pressurization, and the influence of vessel size and fill level on the rate at which wall temperature and pressure increase. Figure 7 shows an example of comparison between CFD results and experimental measurements, considering two tests presented in literature [27] involving a 5-ton LPG tank with two different filling degrees: 22% and 72%. The CFD predictions are in general agreement with the experimental data, given the uncertainty in thermal boundary conditions in the experiments.

In their conclusions, Scarponi et al. proposed that the approach of modelling the pressurization on the basis of the liquid surface temperature can be improved upon by considering the balance of evaporation and condensation occurring at different zones throughout the liquid phase. Figure 8 shows modeling results of evaporation and condensation zones recreating LPG vessel tests [27] of three differing sizes.

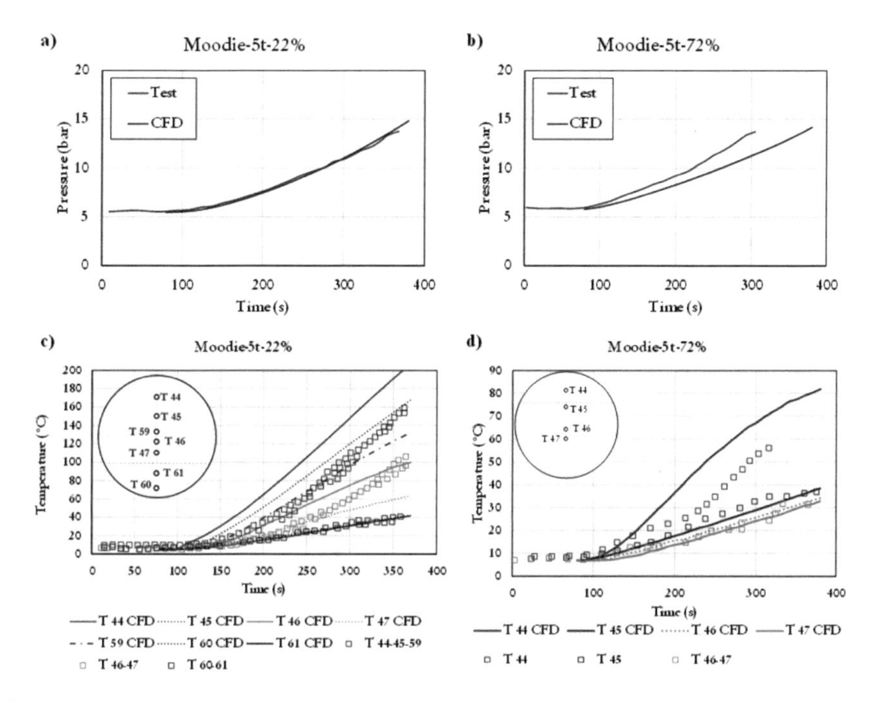

Fig. 7. Comparison of CFD results and experimental measurements of pressure (a and b) and temperature (c and d) transient profiles for tests involving a 5 ton LPG tank [27] with a filling degree of 22% and 72% (adapted from Scarponi et al. [26]).

The study by Scarponi et al. [26] draws attention to the importance of both boiling and compression of vapor due to liquid thermal expansion as mechanisms that can drive rapid pressurization prior to activation of relief devices. Thermal stratification is highlighted as having a proportionately greater influence on the pressurization rate within bigger vessels and at high fill levels.

Despite the progress made in CFD models of LPG vessels there remain numerous challenges to overcome, including:

- simulation of the action of relief devices remains a significant challenge;
- implementation of an improved phase change model with more advanced two-phase flow models that allow separate velocity and temperature fields for the two phases;
- validation against data with well-characterized thermal boundary conditions;
- computational requirements for 3D simulations.

The Euler-Euler multiphase model and the heat flux partitioning model (known as the RPI model) for wall boiling [28] show promise for modelling liquids stored below saturation pressure, however use of the RPI wall boiling model requires knowledge of bubble parameters including the departure diameter, departure frequency and nucleation site density; availability of these parameters is often poor for natural convection scenarios and non-water fluids.

Further research is required to address the current deficiencies in our ability to model the response of pressure vessels to fire. Improvements in modelling capabilities require validation against reliable test evidence, with focus on control of thermal boundary conditions.

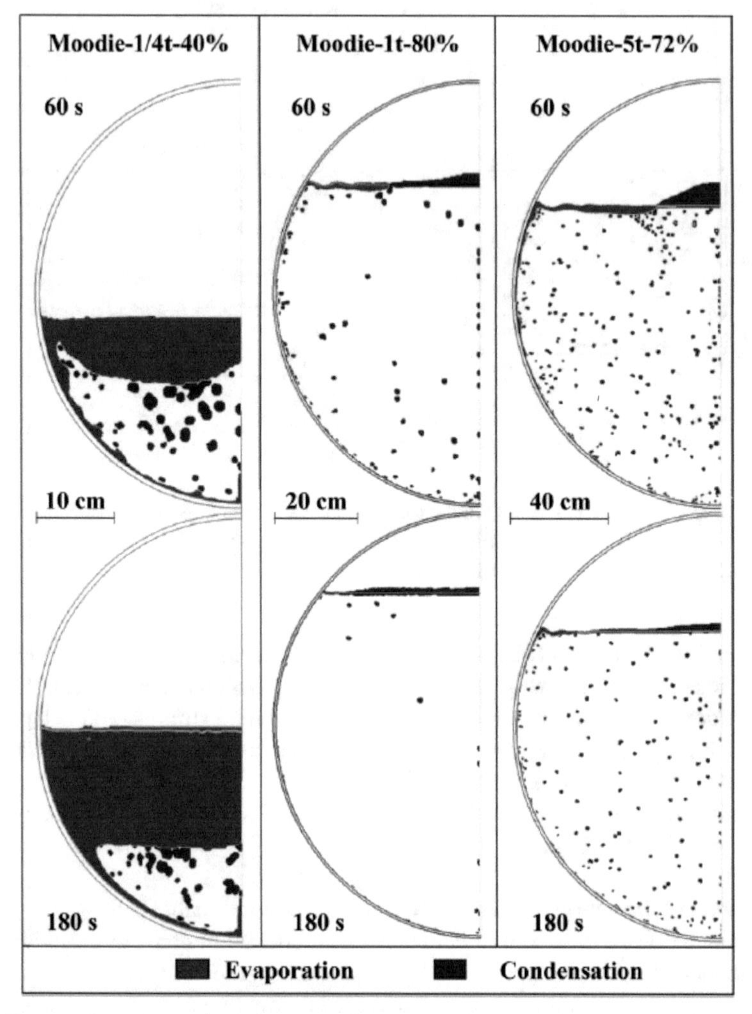

Fig. 8. Evaporation and condensation zone simulations of various sizes of LPG vessel tested by the HSE (adapted from Scarponi et al. [26])

4 Selected Experimental Results

The importance of using appropriate protection measures to prevent vessel catastrophic failure is highlighted by a series of tests performed at BAM [29]. Destructive testing of three vessels was performed, with the intention of keeping constant all parameters except for the means of protection. The three types of protection were as follows:

1. PRV only
2. Thermal coating only
3. PRV and thermal coating

Figure 9 illustrates the results from this comparative study, plotting the progression of maximum wall temperature and vessel pressure with time. Also plotted is a typical fire temperature, which was controlled in the region of 900–1000 °C.

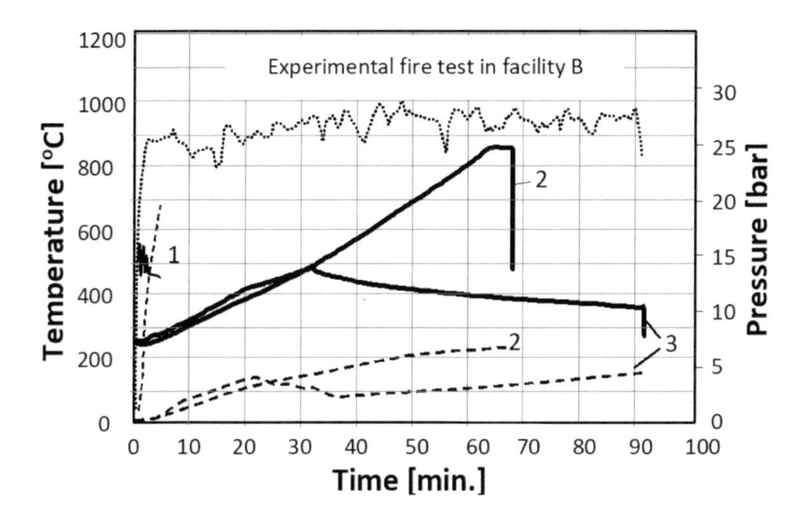

1: Tank with safety valve.
2: Tank with fire protection coating.
3: Tank with fire protection coating and safety valve.

Fig. 9. Temperature and pressure development of vessels with different protection systems

Case 1 (PRV only) gave a rapid temperature and pressure increase, resulting in a BLEVE in approximately 4 min.

The thermal coating used was an epoxy intumescent that reduces the rate of heat transfer to the vessel wall by expanding during fire and forming an insulative char. Case 2 (thermal coating only) showed a significant reduction in the rate of wall temperature rise and pressure rise. A typical activation pressure for a standard PRV fitted to an LPG vessel is approximately 15 bar. This value was reached within a minute when the vessel had no thermal coating but delayed to 30 min when a coating was present. Case 2 exhibited a reasonably linear pressure rise until the test was terminated for safety reasons at close to 70 min.

Case 3 (PRV and thermal coating) exhibited a similar rate of temperature and pressure rise to case 2 until the PRV activated for the first time, after which the pressure decreased for the remainder of the test. The PRV remained partly open for the remaining duration of the test and therefore the pressure decreased slowly despite the maximum wall temperatures starting to increase again from approximately 40 min.

A comparison of the time to failure or termination of all three cases is given in Fig. 10. An important conclusion from these tests is that pressure relief alone may not be sufficient to prevent vessel failure. Use of a thermal coating alone will prolong the time to failure, however failure is still a significant possibility. Use of a thermal coating with pressure relief provides the best protection, capable of protecting a vessel for durations required by regulations, facility risk assessments, or until the vessel empties.

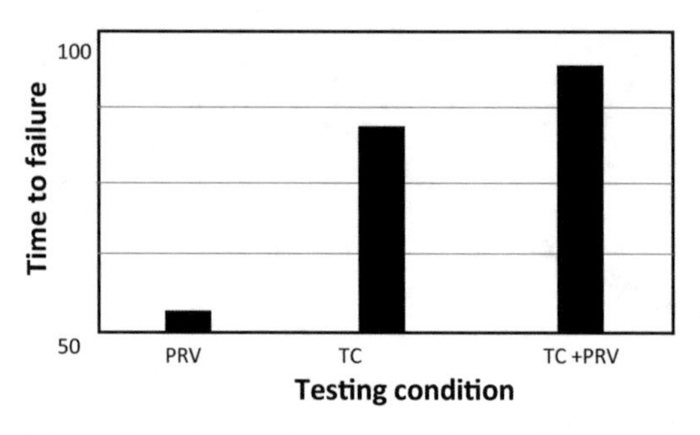

Fig. 10. The effect of thermal coatings and pressure relief on time to failure for the PRV only, the thermal coating (TC) only and both PRV and TC

The results described above can be put into a practical context by including burst pressure calculations in accordance with existing codes. Figure 11 shows a graph of burst pressure against temperature calculated using ANSYS FEA software. This figure illustrates a straight-line relationship, similar to the allowable stress condition according to the relevant American Society of Mechanical Engineers (ASME) code. Values from all three cases described above have been added to the figure, taken at the point of failure or termination. Case 1 can be seen to fail at a temperature and pressure combination that fits the calculation well. Minor discrepancies may reflect undetected localized regions of higher temperature on the vessel between thermocouples or variations in material properties. The two cases with thermal coatings appear to have had a significant margin of safety remaining, however the margin of safety applied to vessels with thermal coating system should take into account the possibility of local defects in the system.

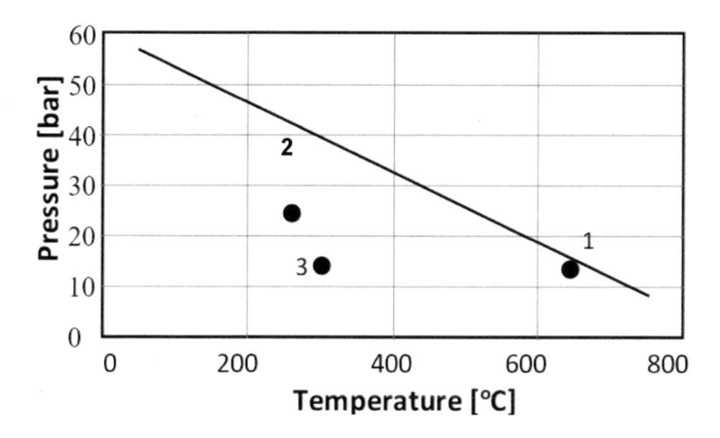

1: PRV (Failure)
2: Thermal coating only (No failure)
3: PRV + Thermal coating (No failure)
——— Burst temperature (ANSYS)

Fig. 11. Test results in the context of regulation and theoretical burst pressure

5 Conclusions

The consequences of pressure vessel exposure to fire can be severe, as demonstrated by fatalities, injuries and evacuations caused by transportation accidents. Fires, caused by either a loss of containment from pressure vessels of flammable substances or by fuel sources associated with the transport of such substances, result in a risk of explosion and escalation of the incident.

Understanding the behavior of the fluid inside the vessel is critical to assessing the probability such an event will occur, and to assessing the severity of such an event if it were to occur. An understanding of the thermohydraulic behavior of the fluid is critical in this regard. The ability of society to model the mechanical response of the vessel wall in circumstances of known temperature and pressure is reasonable, as is knowledge of the thermal boundary conditions in event of well-characterized fires. Limitations on our ability to predict pressure vessel response to fire are primarily practical issues (lack of knowledge of the fire conditions and condition of vessel, given the wide range of possible accident scenarios) and thermohydraulic issues associated with the pressure rise within the vessel.

The limitations of existing single- and multi-zone models have been highlighted. Single-zone models fail to predict the thermal stratification that drives early pressure rise, whereas multi-zone models are superior but rely on empirically derived parameters to govern the relationship between zones.

Improving our ability to model the pressurization rate of vessels is a field of current academic interest. It requires development of improved CFD models to capture the fluid response accurately. Future work should focus on implementing appropriate multiphase flow and phase change models, implementing pressure relief action, and predicting three-dimensional effects.

Future CFD work must be accompanied by experimental work to provide data for validation. Existing data suffers from imprecise control of thermal boundary conditions and future experimental work should focus on control of these. Improved instrumentation to capture temperature and flow within the boundary layers and to capture properties required for implementation of wall-boiling models within CFD is also required.

Existing test facilities at BAM have been described. These range from destructive test facilities capable of directly assessing the influence of changes in boundary conditions (fire type, fire coverage, pressure relief devices and other protection systems) to a thermohydraulic monitoring test rig capable of performing PIV studies inside a vessel during fire exposure. PIV is a method of capturing flow behavior over a wide field of view and is therefore useful for development of validation data for the improved CFD models required.

A summary has been provided of tests which highlight the inadequacy of a PRV alone to reliably prevent catastrophic vessel failure. Tests performed at BAM demonstrate that the combination of a PRV and a thermal protection system is capable of significantly delaying, even preventing, catastrophic failure.

References

1. IMPEL: Derailment of LPG tank-wagons followed by a UVCE explosion and an intense fire. 29 June 2009, Viareggio, Italy. Report 36464, French Ministry for Sustainable Development (2011)
2. ECE: Evaluation of the global and regional impact of UNECE regulations and United Nations Recommendations on the transport of Dangerous Goods (2005–2014). Report. Economic Commission for Europe (2016)
3. BAM: Experience BAM. History of BAM. Bundesanstalt für Materialforschung und – prüfung (2018). https://www.bam.de/Navigation/EN/About-us/Experience-BAM/BAM-History/bam-history.html
4. BAM: TES Technical Safety – Dangerous Goods Containments (2018). https://www.tes.bam.de/en/umschliessungen/behaelter_radioaktive_stoffe/behaelterpruefungen/index.htm
5. Roberts, T.A., Buckland, I., Schirvill, L.C., Lowesmith, B.J., Salater, P.: Design and protection of pressure systems to withstand severe fires. Process Saf. Environ. Prot. **82**(2), 89–96 (2004)
6. Bradley, I., Otremba, F., Birk, A.M., Bisby, L.: Novel Equipment for the Study of Pressure Vessel Response to Fire. International Mechanical Engineering Congress, Phoenix, USA (2016)
7. Bradley, I., Scarponi, G.E., Otremba, F., Cozzani, V., Birk, A.M.: Experimental analysis of a pressurized vessel exposed to fires: an innovative representative scale apparatus. Chem. Eng. Trans. (2017). https://doi.org/10.3303/CET1757045
8. Raffel, M., Willert, C.E., Wereley, S., Kompenhans, J.: Particle Image Velocimetry. A Practical Guide. Springer, Heidelberg (2007)
9. Birk, A., Yoon, K.T.: High-temperature stress-rupture data for the analysis of dangerous goods tank-cars exposed to fire. J. Loss Prev. Process Ind. **19**, 442–451 (2006). https://doi.org/10.1016/j.jlp.2005.11.003
10. Birk, A., Dusserre, G., Heymes, F.: Analysis of a propane sphere BLEVE. Chem. Eng. Trans. **31**, 481–486 (2013). https://doi.org/10.3303/CET1331081

11. Aydemir, N.U., Magapu, V.K., Sousa, A.C.M., Venart, J.E.S.: Thermal response analysis of LPG tanks exposed to fire. J. Hazard. Mater. **20**, 239–262 (1988). https://doi.org/10.1016/0304-3894(88)87015-8
12. Beynon, G.V., Cowley, L.T., Small, L.M., Williams, I.: Fire engulfment of LPG tanks: HEATUP, a predictive model. J. Hazard. Mater. **20**, 227–238 (1988). https://doi.org/10.1016/0304-3894(88)87014-6
13. Birk, A.M.: Experimental investigation of a cylindrical vessel engulfed in fire with a burning relief valve flare present. In: American Society of Mechanical Engineers, Heat Transfer Division (Publication) HTD (1988)
14. Birk, A.M.: Modelling the response of tankers exposed to external fire impingement. J. Hazard. Mater. **20**, 197–225 (1988). https://doi.org/10.1016/0304-3894(88)87013-4
15. Birk, A.M.: Development and validation of a mathematical model of a rail tank-car engulfed in fire. Dev. Valid. a Math. Model a Rail Tank-car Engulfed Fire Ph.D. thesis, Queen's University, Kingston, Ontario, Canada (1983)
16. Dancer, D., Sallet, D.W.: Pressure and temperature response of liquefied gases in containers and pressure vessels which are subjected to accidental heat input. J. Hazard. Mater. **25**, 3–18 (1990)
17. Gong, Y.W., Lin, W.S., Gu, A.Z., Lu, X.S.: A simplified model to predict the thermal response of PLG and its influence on BLEVE. J. Hazard. Mater. **108**, 21–26 (2004). https://doi.org/10.1016/j.jhazmat.2004.01.012
18. Graves, K.W.: Development of a Computer Model for Modeling the Heat Effects on a Tank Car. US Department of Transportation, Federal Railroad Administration, Washington DC (1973)
19. Johnson, M.R.: Tank Car Thermal Analysis, Volume 1, User's Manual for Analysis Program. Department of Transportation, Federal Railroad Administration, Washington DC (1998)
20. Johnson, M.R.: Tank Car Thermal Analysis, Volume 2, Technical Documentation Report for Analysis Program. US Department of Transportation, Federal Railroad Administration, Washington DC (1998)
21. Birk A.M.: Review of AFFTAC Thermal Model. TP 13539 E, Transport Canada (2000)
22. D'Aulisa, A., Tugnoli, A., Cozzani, V., Landucci, G., Birk, A.M.: CFD modeling of LPG vessels under fire exposure conditions. AIChE J. **60**, 4292–4305 (2014)
23. Bi, M.S., Ren, J.J., Zhao, B., Che, W.: Effect of fire engulfment on thermal response of LPG tanks. J. Hazard. Mater. **192**, 874–879 (2011)
24. Hadjisophocleous, G.V., Sousa, A.C.M., Venart, J.E.S.: A study of the effect of the tank diameter on the thermal stratification in LPG tanks subjected to fire engulfment. J. Hazard. Mater. **25**, 19–31 (1990)
25. Birk, A.M., Yoon, K.T.: High Temperature Stress Rupture Testing of Sample Tank-Car Steels, TP 14356E, Transportation Development Centre, Transport Canada (2004)
26. Scarponi, G., Landucci, G., Michael Birk, A., Cozzani, V.: LPG vessels exposed to fire: scale effects on pressure build-up. J. Loss Prev. Process Ind. **56** (2018). https://doi.org/10.1016/j.jlp.2018.09.015
27. Moodie, K., Cowley, L.T., Denny, R.B., Small, L.M., Williams, I.: Fire engulfment tests on a 5 tonne LPG tank. J. Hazard. Mater. **20**, 55–71 (1988)
28. Kurul, N., Podowski, M.Z.: On the modeling of multidimensional effects in boiling channels. In: Proceedings of the 27th National Heat Transfer Conference, Minneapolis, MN, USA (1991)

29. Sklorz, Ch., Otremba, F., Balke, Ch.: Performance of dangerous goods tanks in a fire, Internal Report. BAM. Berlin (2015)
30. Otremba, F., Bradley, I., Romero Navarrete, J.A.: Boiling thermohydraulics within pressurized vessels. In: Lecture Notes in Engineering and Computer Science: Proceedings of The World Congress on Engineering and Computer Science 2018, San Francisco, USA, 23–25 October 2018, pp. 538–542 (2018)

Synthesis of Mass Exchanger Networks Using Sequential Techniques

Kelvin Odafe Yoro[1](✉), Adeniyi Isafiade[2], and Michael Olawale Daramola[1]

[1] School of Chemical and Metallurgical Engineering, Faculty of Engineering and the Built Environment, University of the Witwatersrand, WITS, Johannesburg 2050, South Africa
kelvin.yoroo@gmail.com, michael.daramola@wits.ac.za
[2] Department of Chemical Engineering, University of Cape Town, Rondebosch, Cape Town 7701, South Africa
aj.isafiade@uct.ac.za

Abstract. Mass exchange units are important in many industrial processes. Therefore, there is a need to design and operate these units in a coordinated and integrated manner. This chapter presents a systematic approach for the synthesis of mass exchanger networks involving utility targeting. A sequential technique for targeting external mass separating agent and developing an optimal network design is discussed. The proposed technique was applied to a CO_2 adsorption process involving two mass separating agents (MSAs) that do not overlap. A thermodynamic analysis of the CO_2 adsorption process was outlined using the composition interval method. Feasible structures were formulated, and the synthesis task was expressed in a two-stage targeting procedure as an optimization task. This chapter also discussed a trade-off between the process and external MSA in the case study considered to determine the minimum amount of external MSA required for a typical CO_2 capture process. The results obtained indicated that mass integration via process synthesis is an effective strategy to minimize the quantity of external utilities in industrial processes.

Keywords: Adsorption · Composition interval · Mass exchanger networks · Mass separating · Agents targeting · Network design · Utility targeting

1 Introduction

Mass exchangers are direct contact mass-transfer units that utilize mass separating agents (MSAs) to preferentially separate certain components (example; impurities, pollutants and by-products) from a rich stream to a lean stream. Popular mass exchange operations include adsorption, stripping, absorption, ion exchange, and solvent extraction. Mass Exchange Networks (MENs) are integrated mass exchangers which can be used in the chemical industry to decrease waste generated to tolerable levels at the cheapest cost [1]. Mass exchanger network synthesis (MENS) has been used in process engineering to minimize the quantity of external mass separating agents (MSAs) required in a separation network. A few advances have been reported in approaches for the synthesis of

S.-I. Ao et al. (Eds.): WCECS 2018, *Transactions on Engineering Technologies*, pp. 173–185, 2020.
https://doi.org/10.1007/978-981-15-6848-0_14

mass exchanger networks using techniques like pinch technology and mathematical programming. The pinch technology approaches, which are sequential, use pinch concepts and graphical illustrations to decompose the mass exchanger network design problem into sub-problems to minimize utility costs, number of units and investment costs in an industrial process [2]. The mathematical programming methodologies, which are simultaneous in nature, offer the possibility of synthetizing the network in a single step by formulating the problem as a mixed-integer nonlinear programming (MINLP) model. However, determination of an optimal network is often difficult due to the non-convexity of the mathematical representation of the problem. The pinch technology-based approaches (sequential techniques) on the other hand use pinch concepts and graphical illustrations to decompose the mass exchanger network design problem into sub-problems to minimize energy costs, number of units, investment costs and the amount of material to be consumed in an industrial process [2].

Most of the current researches in process synthesis focus on heat exchanger network synthesis [3, 4]. The few literature reports on mass exchanger network synthesis mostly discussed single-component problems using simultaneous approaches. Less attention has been dedicated to mass exchanger network synthesis using the sequential techniques. Figure 1 shows a typical mass exchanger with stream flow in counter current directions.

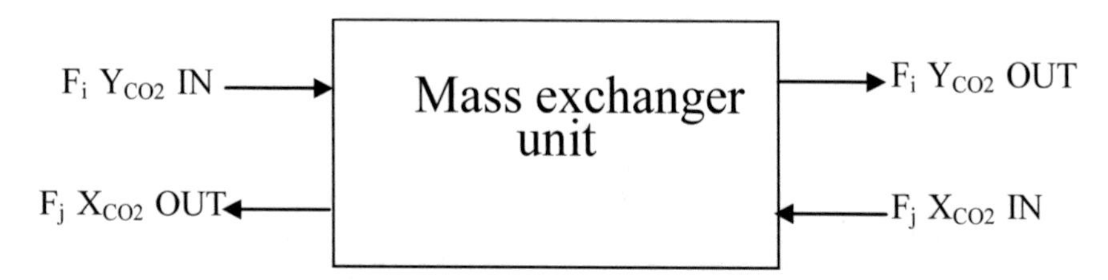

Fig. 1. A typical mass exchanger for counter current CO_2 adsorption

For a typical mass exchanger network, the streams from which the targeted components are removed are designated as rich streams while the streams to which the targeted components are transferred are referred to as the lean-stream (MSA). It is expected that the lean stream should be partially or completely immiscible with the rich stream.

2 Advances in Synthesis of Mass Exchanger Networks

Literature is rich in non-sequentially based methods and approaches for synthesis of mass exchanger networks [5–14]. For instance, the synthesis of separation sequences for mass exchanger networks was first reported by Siirola et al. [15] using heuristic approach. This was the first attempt to synthesize mass exchanger networks for separation systems reported by researchers in the early 70's. The authors established that it might always not be possible to apply heuristic approach to most systems due to thermodynamic limitations. In view of this, Thompson and King [16] introduced a technique that combined heuristics with algorithmic programming to determine types of separation processes alongside their sequences. In another study, an evolutionary approach for the synthesis of multicomponent separation sequences involving mass exchanger networks was proposed by Stephanopoulos and Westerberg [17]. This evolutionary approach was further developed into a combined heuristic and evolutionary strategy for the synthesis of mass exchanger networks. El-Halwagi and Manousiouthakis [18] proposed a procedure for optimal synthesis of MENS using sequential method (pinch analysis). The authors proposed that, by using a composition interval method, a composition pinch point could be set on the mass transfer composite curves and the minimum allowable composition difference 'ε' can be ascertained. The minimum utility targets (minimum MSA consumed) can be obtained from the mass transfer composite curves before any network design using this approach. The use of P-graph theory to solve problems involving mass exchanger networks (MENS) was introduced by Friedler et al. [19]. The work of Friedler et al. [19] was later improved by combining the P-graph theory with non-linear mathematical programs of Cabezas et al. [20] to determine the network structure and its operating conditions. It is worthy to note that most of the methods highlighted applied mathematical programming (simultaneous techniques). The main disadvantage of the simultaneous methods is the difficulty faced while setting up and understanding the mathematical formulations involved. In response to this, a pinch-based methodology for synthesis of mass exchanger networks (MENs) considering the capital cost target was presented by Hallale and Fraser [21]. The authors addressed a special case involving water minimization in which both utility and the capital costs were targeted before network design. Application of the principles of Pinch Technology for mass integration has been adequately reported by El-Halwagi and Manousiouthakis [18], but it did not focus on low temperature studies like CO_2 adsorption. These two pinch technology based methods have shown that principles of pinch technology are essential in providing mass targets for a mass exchange network.

In this chapter, pinch concepts are used to synthesize an optimal network of mass exchangers for optimal adsorption of CO_2 from a rich flue gas stream, with 'Polyaspartamide' (adsorbent) as the mass separating agent that must be minimized, while packed-bed adsorption column is the mass exchange unit. In the sequential technique proposed in this chapter, the overall synthesis problem is decomposed within a sequence of smaller problems that are easier to solve. However, this approach cannot explore the interactions between sub-systems to obtain an improved solution. The sequential procedure in this study can further be used to target optimal loads and levels for multiple utilities by using the process grand composite curve on a direct numerically-based technique. It can also be used to determine the consumption target of each utility by maximizing the use of the cheap utilities (process MSA) and minimizing the loads of expensive ones (external MSA) that can be used to remove CO_2 from a CO_2-rich process stream with little operating costs.

3 Problem Statement

The problem addressed in this chapter involves a preferential adsorption of CO_2 from a CO_2/N_2 mixture using a solid sorbent, and is expressed as follows; 'Given are a set of rich streams 'R' and a set of lean streams 'S', where the rich streams have flowrate represented as 'F', and the lean streams have flowrate represented as 'L'. The rich stream compositions are to be reduced from supply compositions Y_i to target compositions Y_t, while the lean streams have supply compositions X_i increasing to target compositions X_t. The flow rates of the lean streams are not known for the process MSA, it is restrained by a definite maximum flow rate expressed in Eq. (1).

$$L_j \leq L_L \quad (1)$$

The task is to synthesize an optimal network of mass exchanger units that can preferentially separate CO_2 'C' from the rich mixture of CO_2 and N_2 using a set of mass separating agents, at a minimum cost. The process MSA is virtually free and already exists in the adsorption plant site, while the external MSA (S_2) can be purchased.

To solve the problem in this study, the following major assumptions have been made;

1. The flow pattern in the mass exchanger unit is counter current.
2. Throughout the network, the mass flow rate of each stream is constant.
3. To attain a single equilibrium relation, temperature and pressure do not change throughout the network of each stream.
4. Exchange of mass between rich-rich and lean-lean streams is not allowed.

4 The Synthesis Procedure

The synthesis task for the problem highlighted in Sect. 3 is presented as a two-stage task with composition difference as the driving force. First, the minimum amount of mass separating agents for CO_2 adsorption was determined using thermodynamic analysis before designing the network configuration. Flow rates of the external MSA were then determined and potential mass exchange operations screened according to the procedure of Kaguei and Wakao [22]. In the next stage of the synthesis task, a composition interval for each feasible pair of rich-lean stream is used in a generalized procedure to minimize the total annualized cost of the MENS. A case study involving CO_2 adsorption was used to test the application of MENS to a typical CO_2 capture problem. Minimum utility target networks were generated by merging any rich-end design with any corresponding lean-end design. Other possible results obtainable from using the proposed pinch-based method described in this section are presented in Fig. 2.

The concept of pinch analysis can also be used to formulate a composition interval table which could be used to locate pinch bottlenecks (pinch points) at specified minimum allowable composition difference ($\varepsilon = 0.0001$). This value of ε was chosen to avoid infinite sizes of mass exchanger which could make the generated network design expensive. The minimum utility requirement of the capture process was also determined using the same approach. Equilibrium equation for the transferable component (CO_2) is in a linear relationship between the process MSAs and the rich stream concentration as expressed in Eq. (2).

$$Y = mX^* + b \quad (2)$$

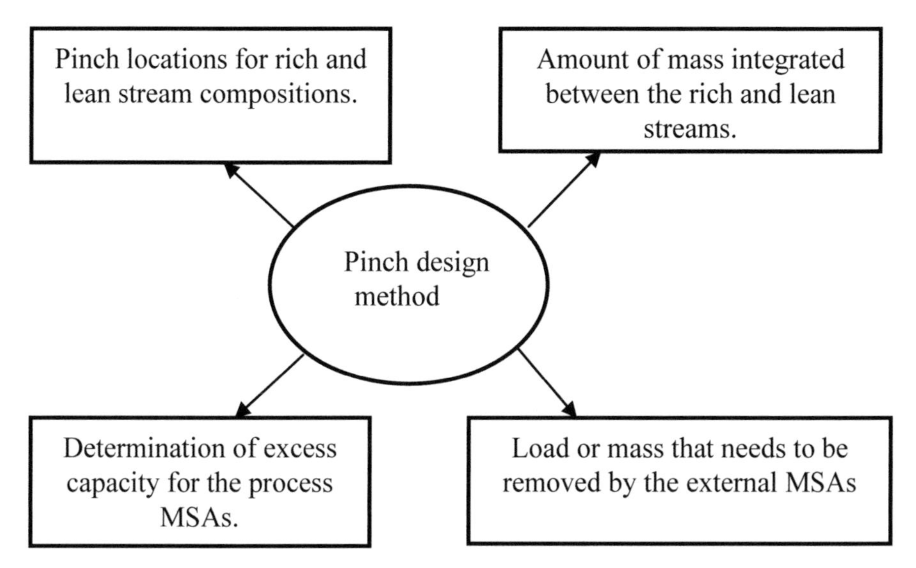

Fig. 2. Possible results from using pinch-based methods.

X^* is the maximum equilibrium composition of the lean stream theoretically attainable. The minimum allowable composition difference 'ε', if included in the equilibrium relation expressed in Eq. (2), can be used to avoid infinite sizes of mass exchangers. The new equilibrium equation is presented in Eq. (3).

$$Y = m(X^* + \varepsilon) + b \tag{3}$$

The corresponding composition scales of components in the lean and rich streams can be determined from Eq. (3). In the composition interval table presented in this study, composition intervals correspond to the supply or target composition of components in each stream. The material balance on CO_2 that is moved from stream i to stream j is obtainable using Eq. (4).

$$F_i(Y_{CO2}IN - Y_{CO2}OUT) = L_j(X_{CO2}OUT - X_{CO2}IN) \tag{4}$$

Y_{CO2} IN and Y_{CO2} OUT are the inlet and outlet compositions of CO_2 in the rich stream, X_{CO2} OUT and X_{CO2} IN are the outlet and inlet compositions of CO_2 in the lean stream respectively. The following points should be noted in the synthesis task;

1. The type of mass-exchange operation that should be used (e.g., absorption, adsorption).
2. Mass separating agent that should be selected (e.g. solvents, adsorbents).
3. The optimal flowrate of each MSA
4. How these MSA's should be matched with the rich streams (i.e. stream parings) and the optimal system configuration.

The composition interval diagram for both rich and lean stream is shown in Table 1. Excess capacity of the external MSAs needed for the capture process was obtained from the composition interval table while the molar flow rates of the lean and rich streams are shown in Table 2. The composition interval diagrams (CID) constructed in this chapter was used to calculate the minimum mass flow rates of MSA (adsorbent) required for the capture process.

5 Minimum Number of Units Targeting in MENS

Detailed technique for targeting the minimum number of units for mass exchanger networks have been explained by Foo et al. [6]. Just as is the case with heat exchanger networks, fewer units of mass exchangers in a network minimize its complexity. In addition, fewer units of mass exchangers also lead to reduced pipework, foundation, maintenance and instrumentation. Minimum number of units in a MENS network 'U' can be determined by linking it to the total number of streams in a network, according to the Eq. (5).

$$U = NR_S + NS_L - NS_N \tag{5}$$

Where NR_S is the number of rich streams, NS_L is the number of lean streams and NS_N is the number of independent sub-networks into which the original network can be subdivided. It is worth stating that owing to the presence of the pinch composition which divides the problem into two distinct sub-networks, Eq. (5) could be applied independently above and below the pinch. If the case to be considered has different time intervals, process rich or lean streams may exist in more than one time interval. To reduce the number of units, mass exchangers connecting the same pair of rich and lean streams are normally re-used in each time interval. In view of this, it is recommended that the targeting technique should consider the opportunities to reuse these exchangers. This is tested for CO_2 adsorption studies in Examples (1). The value of the minimum approach composition chosen in the case study will prevent the infinite sizes of the mass exchangers during design. This is very vital because it reduces the overall design cost of the CO_2 capture process. The reason adsorption of CO_2 was considered in this paper is to capture CO_2 from a mixture of gases. According to Yoro et al. [23–25], the removal of CO_2 is necessary because CO_2 is a greenhouse gas that has negative impact on the atmosphere if allowed to be emitted. Since the example considered in this chapter is a continuous process, Eq. 5 can be used to determine the minimum number of units. But if it is a batch process with time intervals involved, Eq. (6) can be used with a time-dependent composition interval table to calculate the minimum number of units for the mass exchangers.

$$U_k = NR_{S,k} + NS_{L,k} - NS_{N,k} \tag{6}$$

The streams actually exist in short time intervals in a batch process. Hence, we suggest that if the same pair of streams exist in more than one time interval, the numbers of "additional exchangers", U_{AE} can be expressed from Eq. (6) as;

$$U_{AE} = N_{TI} - 1 \tag{7}$$

Where N_{TI} is the number of time intervals where both streams co-exist. Hence, for a mass exchanger network with 'j' additional exchanger, the minimum number of units is given by Eq. (8).

$$U = \sum U_k - \sum U_{AE}, j \tag{8}$$

Example 1
This example considers two rich and two lean streams. It is adapted and modified from Yoro [23] which involves the preferential adsorption of CO_2 onto polyaspartamide from two gas streams. The gas compositions given in Yoro [23] was modified in this study by making appropriate molar and mass flow rate conversions where necessary. The problem data for this example is presented in Tables 1 and 2. Supply composition of CO_2 in the gas mixture is 15 wt%. Two mass separating agents (process and external MSAs) were considered with a minimum approach composition specified as 0.0001.

Table 1. Rich and lean streams data for example 1

Rich streams	F	Y_{in}	Y_{out}	Density
	(kmol/h)	(kmol/kmol)	(kmol/kmol)	(kg/m^3)
R1	1.51	0.15	0.06	1.98
R2	1.23	0.15	0.08	1.98
Lean streams	K	X_{in}	X_{out}	Density
	(kmol/h)	(kmol/kmol)	(kmol/kmol)	(kg/m^3)
S1	1.33	0.10	0.05	1.25
S2	∞	0.10	0.07	1.25

Since the minimum allowable composition difference 'ε' is an optimizable parameter, it is possible to transfer CO_2 from a rich stream Y to a lean stream X using Eq. (9).

$$X_{CO2,\, j} = \frac{Y_{CO2,\, I} - b_{CO2}}{m_{CO2}} - +_{CO2} \tag{9}$$

Hallale and Fraseer [26] conducted a supertargeting work for mass exchangers and established that the flow of mass separating agents (sorbents) used will be at a minimum value when $\varepsilon = 0.0001$. Hence, the value of ε is adopted in this study as 0.0001. But when ε_{CO2} is increased, operating cost of the capture process increases. Considering the assumptions highlighted for this study, Eq. (5) can be used to create the corresponding composition scales for CO_2 in the lean process streams S_1 and S_2. Data on Table 2 represent the composition mass flows of CO_2 available for capture while the intervals correspond to the head and tail for Example 1. It is assumed here that each lean process stream leaves the network at a specified outlet composition constrained by Eq. (1).

Table 2. Composition intervals for CO_2 adsorption

Intervals	Rich stream Y	Lean stream X	A Input (kg/s)	B Output (kg/s)	C Excess (kg/s)
	0.150 R1	0.130			
1			0.000	0.070	0.070
	0.147 R2	0.110	0.070	0.085	0.015
2					
	0.144	0.080	0.085	0.098	0.013
3					
	0.142	0.060	0.098	0.100	0.002
4					
	0.140	0.050	0.100	0.170	0.070
5			S1		
	0.139	0.040	0.170	0.188	0.018
6					
	0.139	0.000	0.188	-	-
				S2	

The composition interval table comprises 3 columns labeled A, B and C excluding the stream representations. Column A and B represent the input and output flows of CO_2 in the streams while column C represents the excess mass of CO_2 available for capture in each interval. The process stream is represented with an arrow on the composition interval table. The rich process streams are denoted as R1 and R2 while lean process streams are S1 and S2. The target composition of the rich stream or the constrained outlet composition of the lean stream is represented by the arrow head while the tail corresponds to the supply composition of respective streams. For Example 1, the minimum utility network of mass exchangers required by using Eq. (5) is:

$$U = (2 + 1) - (1 - 1)$$

$$U = 3.$$

It is noted from literature that it is thermodynamically infeasible to transfer a negative output of mass between intervals in a composition interval table [18]. Negative mass flows in the columns (if any) must be modified by increasing the input to the first interval. In the data presented on Table 2, no negative mass was recorded. This means that example 1 is thermodynamically feasible without further modification. Since the thermodynamic constraints are guaranteed to be satisfied from the data in Table 2, it is possible to transfer CO_2 from rich to lean streams. In column C, it is noted that the excess CO_2 removal capacity is 0.070 kg/s. Based on the guaranteed thermodynamic constraints in example 1, this excess can be used to remove CO_2 from the rich streams.

The last value in column C (0.018 kg/s) is the minimum mass of CO_2 that is capturable by the external mass separating agent (polyaspartamide) from which the minimum flow is determined. This also implies that for the CO_2 adsorption study in Yoro [23] which lasted for 1200 s adsorption time, about 21.6 kg of CO_2 could be captured per gramme of adsorbent using this method. Additional important information that could be obtained from the composition interval table presented here is the pinch point (a point at which the mass flow of CO_2 available for capture vanishes). From the information provided on Table 2, the pinch point exists between interval 5 and 6 with a minimum mass flow of 0.170 kg/s. This interval represents the most constrained region of the mass exchanger network design because all matches between the rich and lean streams will be subject to a minimum allowable composition difference at the pinch point.

6 Network Design

The approach used for the design of mass exchanger networks in this chapter is based on a modification from the work of Foo et al. [6] and El-Halwagi and Manousiouthakis [18]. Although the work of Foo et al. [6] was based on a batch process (with time intervals), the technique was modified in this work to consider a continuous process without time interval. From the composition interval data on Table 2, it is possible to generate a network of mass exchangers that use minimum external MSA by merging any rich-end design with a corresponding lean-end design. Figure 4 shows a network for minimum utility targets obtained by combining rich and lean-end designs. Figure 4 was obtained sequentially but according to El-Halwagi and Manousiouthakis [18], when a larger number of streams are involved in minimum utility mass exchanger networks, it will be easier to use a simultaneous approach. The minimum utility network in this study involved an extra unit more than the target number of units because of the existence of a pinch which decomposed the task into two distinct problems. A rich end network design is first obtained and shown in Fig. 3 while a lean-end design network is shown in Fig. 4.

Minimum utility targets can be obtained from the composition interval data by combining the rich-end design in Fig. 3 with a lean-end design in Fig. 4. But Linnhoff et al. [27] observed that after merging rich with lean end design for a heat exchanger network, the minimum-utility pinched network subsequently generated will have one more than the target minimum number of exchanger units. Due to the existence of a pinch which decomposes the task into two distinct sub tasks, the minimum number of units compatible with a minimum-utility design would be obtained by applying the general mathematical expression for minimum number of mass exchangers compatible with a minimum-utility design which is presented in Eq. (5).

Therefore, three mass exchangers are required for the synthesis task in this work. Since any minimum-utility network involves one unit more than the target minimum number of units, it is necessary to develop a method for the systematic reduction in the number of units. This proposed method would involve the use of "mass-load loops" and "mass-load paths as suggested by El-Halwagi and Manousiouthakis [18]. The sequential approach presented in this study is straightforward and can be used to determine the CO_2 avoided, CO_2 capturable and the minimum amount of MSA required for the CO_2 capture process in a detailed study. Nonetheless, it is worthy to note that when larger numbers

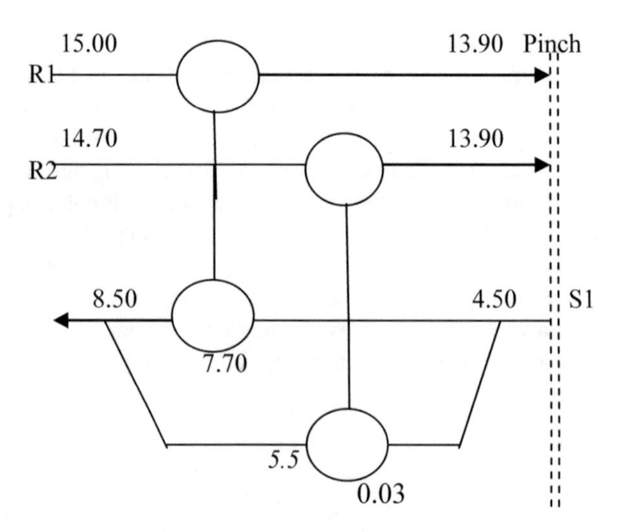

Fig. 3. A rich-end network design

of streams are involved in minimum utility mass exchanger networks, a simultaneous approach is preferred to a sequential approach. The minimum utility network obtained by merging the rich and lean end designs in this study is presented in Fig. 5.

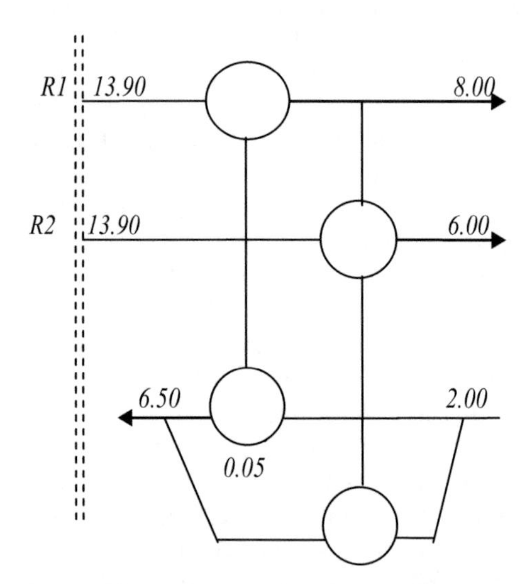

Fig. 4. A lean-end design network

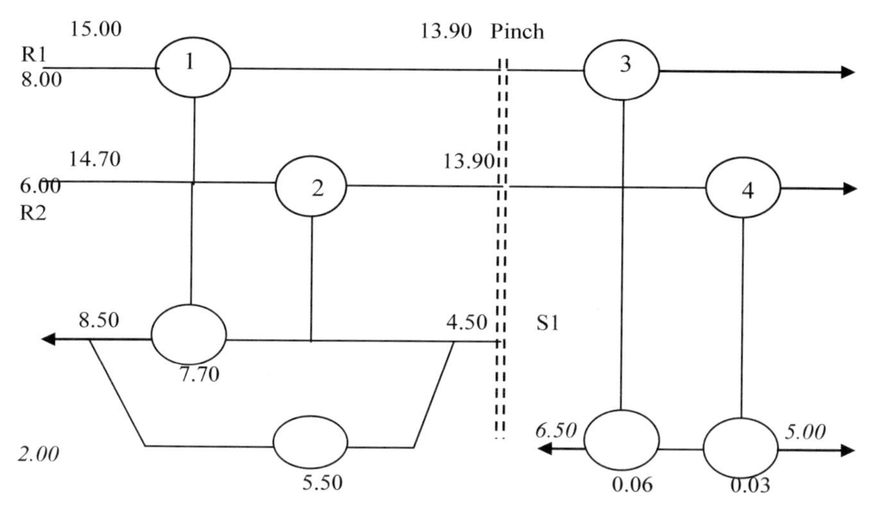

Fig. 5. A minimum-utility network for the adsorption of CO_2 from a CO_2/N_2 gas mixture.

7 Conclusion and Future Outlook

This chapter has shown that minimum utility targets can be achieved for industrial processes by combining rich and lean-end designs. A systematic procedure has been presented to synthesize an optimal mass exchanger network for resource minimization in most industrial processes; this could be modelled further into a non-linear program to optimize the total annualized cost for the CO_2 capture process in a simultaneous approach. The composition intervals presented in this chapter offers a two-feasibility criterion that identifies the essential pinch matches and a stream splitting rationale from the case study problem. The most constrained region (pinch point) of the mass exchanger network design in this study was between the 5th and 6th interval. Although this chapter focused mainly on packed bed adsorption columns, the concepts can be readily applied to other stage-wise adsorption and absorption systems with or without time intervals. However, the limitation of the sequential procedure used in this study is that it cannot consider forbidden stream matches or disallowed mass flows. Future work could consider studying a network with detailed mass exchanger design alongside its capital and operating costs targets.

Acknowledgment. This work was supported in full by the National Research Foundation of South Africa under grant number 107867.

References

1. Gerrard, A., Fraga, E.: Mass exchange network synthesis using genetic algorithms. Comput. Chem. Eng. **22**, 1837–1850 (1998)

2. Yoro, K.O., Isafiade, A.J., Daramola, M.O.: Sequential synthesis of mass exchanger networks for CO_2 capture. In: Proceedings of The World Congress on Engineering and Computer Science 2018. Lecture Notes in Engineering and Computer Science, 23–25 October 2018, San Francisco, USA, pp. 503-508 (2018)
3. Hafizan, A.M., Wan-Alwi, S.R., Manan, Z.A., Klemeš, J.J.: Optimal heat exchanger network Synthesis with operability and safety considerations. Clean Technol. Environ. Policy **18**, 2381–2400 (2016)
4. Pavão, L.V., Costa, C.B.B., Ravagnani, M.A.S.S.: Heat Exchanger Network Synthesis without stream splits using parallelized and simplified Simulated Annealing and Particle Swarm Optimization. Chem. Eng. Sci. **158**, 96–107 (2017)
5. Liu, L., Du, J., El-Halwagi, M.M., Ponce-Ortega, J.M., Yao, P.: A systematic approach for synthesizing combined mass and heat exchange networks. Comput. Chem. Eng. **53**, 1–13 (2013)
6. Foo, C.Y., Manan, Z.A., Yunus, R.M., Aziz, R.A.: Synthesis of mass exchange network for batch processes—part I: utility targeting. Chem. Eng. Sci. **59**(5), 1009–1026 (2004)
7. Isafiade, A.J., Short, M., Kravanja, Z., Moller, K.: Synthesis of mass exchange networks using mathematical programming and detailed cost functions. In: Kravanja, Z., Bogataj, M. (eds.) Computer Aided Chemical Engineering, vol. 38, pp. 1875–1880. Elsevier (2016)
8. Foo, D.C.Y., Tan, R.R.: A review on process integration techniques for carbon emissions and environmental footprint problems. Process Saf. Environ. Prot. **103**, 291–307 (2016)
9. Patole, M., Bandyopadhyay, S., Foo, D.C.Y., Tan, R.R.: Energy sector planning using multiple-index pinch analysis. Clean Technol. Environ. Policy **19**(7), 1967–1975 (2017)
10. Kim, M., et al.: Greenhouse emission pinch analysis (GEPA) for evaluation of emission reduction strategies. Clean Technol. Environ. Policy **18**(5), 1381–1389 (2016)
11. Short, M., Isafiade, A.J., Biegler, L.T., Kravanja, Z.: Synthesis of mass exchanger networks in a two-step hybrid optimization strategy. Chem. Eng. Sci. **178**, 118–135 (2018)
12. Liu, L., Du, J., El-Halwagi, M.M., Ponce-Ortega, J.M., Yao, P.: Synthesis of multi-component mass-exchange networks. Chin. J. Chem. Eng. **21**(4), 376–381 (2013)
13. Velázquez-Guevara, M.Á., et al.: Synthesis of mass exchange networks: a novel mathematical programming approach. Comput. Chem. Eng. **115**, 226–232 (2018)
14. Singh, D., Khanam, S.: Synthesis of mass exchanger network considering piping and pumping costs using process integration principles. J. Chem. Eng. Process Technol. **8**(1), 1–15 (2017)
15. Siirola, J.J., Powers, G.J., Rudd, D.F.: Synthesis of system designs: III. Toward a process concept generator. AIChE J. **17**(3), 677–682 (1971)
16. Thompson, R.W., King, C.J.: Systematic synthesis of separation schemes. AIChE J. **18**(5), 941–948 (1972)
17. Stephanopoulos, G., Westerberg, A.W.: Studies in process synthesis—II: evolutionary synthesis of optimal process flow sheets. Chem. Eng. Sci. **31**(3), 195–204 (1976)
18. El-Halwagi, M.M., Manousiouthakis, V.: Synthesis of mass exchange networks. AIChE J. **35**(8), 1233–1244 (1989)
19. Friedler, F., Varga, J.B., Feher, E., Fan, L.T.: Combinatorial accelerated branch-and-bound method for solving the MIP model of process network synthesis. In: Floudas, C.A., Pardalos, P.M. (eds.) State of the Art in Global Optimization, pp. 609–626. Boston, Kluwer Academic Publishers (1996)
20. Cabezas, H., Argoti, A., Friedler, F., Mizsey, P., Pimentel, J.: Design and engineering of sustainable process systems and supply chains by the P-graph framework. Environ. Prog. Sustain. Energy **37**(2), 624–636 (2018)
21. Hallale, N., Fraser, D.M.: Capital cost targets for mass exchange networks a special case: water minimization. Chem. Eng. Sci. **53**(2), 293–313 (1998)

22. Kaguei, S., Wakao, N.: Relationships between surface diffusivity and pore diffusivity in batch adsorption: measurements of the diffusivities for n-hexane and n-decane in 5 Å molecular sieves. Chem. Eng. Sci. **47**(8), 2109–2113 (1992)
23. Yoro, K.O.: Numerical simulation of CO_2 adsorption behavior of polyaspartamide adsorbent for post-combustion CO_2 capture. M.Sc. thesis, University of the Witwatersrand, Johannesburg, South Africa (2017)
24. Yoro, K.O., Singo, M., Mulopo, J.L., Daramola, M.O.: Modelling and experimental study of the CO_2 adsorption behaviour of polyaspartamide as an adsorbent during post-combustion CO_2 capture. Energy Procedia **114**, 1643–1664 (2017)
25. Yoro, K.O., Amosa, M.K., Sekoai, P.T., Mulopo, J., Daramola, M.O.: Diffusion mechanism and effect of mass transfer limitation during the adsorption of CO2 by polyaspartamide in a packed-bed unit. Int. J. Sustain. Eng. (2019). https://doi.org/10.1080/19397038.2019.1592261
26. Hallale, N., Fraser, D.M.: Supertargeting for mass exchange networks: part II: applications. Trans IChemE **78**, 208–216 (2000)
27. Linnhoff, B., Mason, D.R., Wardle, I.: Understanding heat exchanger networks. Comput. Chem. Eng. **3**(1), 295–302 (1979)

Synthesis and Process Variables Optimization of Al_2O_3-CaO Catalyst (from Eggshell) for Biodiesel Production

Eizabeth Jumoke Eterigho[1](✉), Monica Alueshima Baaki[1], and Silver Eyenbi Ejejigbe[2]

[1] Chemical Engineering Department, Federal University of Technology, P.M.B. 65, Minna, Niger State, Nigeria
jummyeterigho@futminna.edu.ng, baakimonica@gmail.com

[2] 8 Elem close off Rumuibekwe Road, Rumurolu, Port Harcourt, River State, Nigeria

Abstract. The synthesis of alumina-calcium oxide catalyst, optimization of process variables and application of the synthesized catalyst is presented. The effect of the variables during the synthesis of the catalyst was investigated through transesterification. The synthesised catalysts were characterised. Using a 2^4 factorial design, the impregnation ratio, impregnation time, calcination temperature and calcination time as parameters used in catalyst synthesis were investigated on the yield of biodiesel from waste cooking oil. The optimization study showed that; impregnation ratio of 1.17:1 at 1.32 h of impregnation and calcination temperature of 740.31 °C for 2.62 h gave the best results of 75.4% with 95% confidence level when used biodiesel production. The conditions for transesterification of the waste cooking oil were methanol to oil molar ratio of 12:1, 6 wt% catalyst, 60 °C reaction time, and mixing speed of 250 rpm for 90 min. The optimized result agreed with the experimental yield of 76.02% biodiesel using the same conditions.

Keywords: Alumina-calcium oxide · Biodiesel · Optimization · Process parameters · Shell · Yield

1 Introduction

Majorly, homogeneous acid or alkaline based catalysts, such as sulphuric acid and sodium hydroxide are the catalysts used in the production of biodiesel via transesterification [13]. The advantage of this catalyst is that the ester yield is high and the reaction occurs at a comparatively fast rate. But in most cases, the resulting cost of production using homogeneous catalyst is high because of the need to wash off the soap formed during the production process, thereby increasing the total production cost and limiting biodiesel commercialisation [4]. In recent times, locally sourced heterogeneous catalysts have gotten so much attention in the biodiesel production process because of their reusability, environmental friendliness, low cost, high quality yield and efficiency in transesterification of triglycerides when compared to their homogeneous counterparts. Of greater interest is the fact that some of these heterogeneous catalysts could be sourced

S.-I. Ao et al. (Eds.): WCECS 2018, *Transactions on Engineering Technologies*, pp. 186–194, 2020.
https://doi.org/10.1007/978-981-15-6848-0_15

from both domestic and industrial wastes, as well as biological resources, [11]. Previously, these materials were regarded as waste and discarded. But with several research and developments in catalysis and biodiesel production, they have been reported to be of great value. This is due to the advantages inherent in these materials as catalysts. The heterogeneous catalysts are biodegradable, reusable and have the activity to produce high quality biodiesel with high yield [4]. They are regarded as green catalysts because of their environmental acceptability. The use of locally sourced heterogeneous catalysts is economically sustainable because they occur in abundance in our environment [15]. Using these bio-materials as catalysts eliminates the problem of disposal. Research has presented these locally sourced heterogeneous catalysts as good alternatives to the synthetic heterogeneous catalysts, particularly, in biodiesel production [12].

Calcium oxide is an example of a heterogeneous catalyst that is abundantly and cheaply available in the environment. It occurs naturally in animal bones, and shells such as eggshells, oyster shells, and snail shells. According to [4], eggshells are a rich source of calcium oxide. However, researchers have encountered the problem of catalyst leaching with the use of calcium oxide [3, 9]. Several authors have reported that after a few cycles, the solid catalyst leach into the reacting medium thereby, reducing the quantity and quality of the ester produced (Eterigho *et al.* 2018). In order to curb this challenge, several supports have been developed upon which these solid catalysts are anchored inorder to reduce leaching and increase the number of times they can be reused for a reaction. Anchoring the catalyst on a support enhances the value of the catalyst and reduces the risk associated with it. Catalyst supports are of different materials; silica, alumina, potassium hydroxide and chromium have been used by different researchers. A study by [16] presents gamma-alumina as a good support for eggshell derived calcium oxide.

The alumina used in this work was sourced from kaolinite clay, Kutigi, in Nigeria. Calcium oxide supported with alumina has been found to have improved catalytic properties that permit good reactions of the reactants [8]. In view of these, this research attempts to synthesised calcium oxide from eggshell supported on alumina as catalyst with high activity by optimizing the process parameters. The alumina was de-aluminated from locally sourced clay. The activity of the synthesised catalyst was tested in the production of biodiesel using waste cooking oil.

2 Experimental

Kaolinite clay was gotten from Kutigi in Niger State, Nigeria. Alumina was extracted from the clay using hydrochloric acid by leaching [1]. The waste chicken eggshells were collected from a local restaurant; washed thoroughly, dried at 100 °C for 12 h, grinded to 125 μm mesh size and calcined at 900 °C for 4 h. The calcium oxide was dissolved in 50 mL of distilled water containing 10 g of alumina. The ratios of the calcium oxide to alumina was 1:1 and 2:1. The slurry was stirred for a period of 1 and 2 h respectively at room temperature using a magnetic stirrer. After which the slurry was oven dried at 100 °C for 12 h and the dry sample was calcined at temperatures of 650 °C and 750 °C for 2 and 3 h. The impregnation ratio, time, calcination temperature and calcination time were varied using 24 factorial design. The catalyst was characterized using FT-IR, XRF,

BET, and TGA, [16]. 100 mL of the waste cooking oil was pre-treated before use. The treated oil was transferred into a conical flask and heated to a temperature of 60 °C. A mixture of concentrated H_2SO_4 (1% w/w) with methanol (30% v/v) was separately heated at 60 °C and added to the heated oil in the flask. The mixture was stirred for 1 h and allowed to settle for 2 h in a separating funnel. The clean oil was then withdrawn. 30 mL of pre-treated waste cooking oil was poured into a conical flask and heated in a water bath until the oil attained a temperature of 60 °C. 50 mL of methanol and 6 wt% catalyst was added to the oil and the mixture was returned back to the water bath at 60 °C and mixture stirred as 250 rpm for 90 min. The methanol to oil ratio was 12:1 [16]. The conical flask was equipped with a stopper to prevent the escape of methanol by evaporation. After 90 min the mixture was removed and the catalyst was separated by filtration. The liquid was then decanted into a separating funnel and left overnight for separation of the biodiesel from the glycerol by gravity. The glycerol which formed the lower layer was collected and the biodiesel yield was determined [7]. The reusability of the catalyst was tested for 5 runs under the same conditions of reaction.

3 Results and Discussion

The X-ray fluorescence carried out on the CaO/Al_2O_3 catalyst showed that CaO and alumina were present as major components in the catalyst in nearly equal amounts. While other oxides like silicon oxide, zinc oxide, titanium oxide, and iron oxide were present in very minute quantities as impurities, the alumina and calcium oxide have the largest percentage compositions of 46.163% and 45.472% respectively. The result reflects a 1:1 ratio of impregnation used in preparing the catalyst. For the FTIR, in the O–H stretching region, the sample shows prominent bands at 3672–3996 cm^{-1} corresponds to Al–OH stretching. Inner hydroxyl groups, lying between the tetrahedral and octahedral sheets, give the absorption at 3672 cm^{-1}. In the bending region mode, the catalyst shows series of IR bands with peak maxima at 1065, 1427 and 1512 cm^{-1}. The peak at 1512 cm^{-1} is quite intense and could be attributed to the bending vibration mode of physisorbed water on the surface of catalyst produced due to leaching; while the 1065 and 1427 cm^{-1} bands is due to the stretching vibrations of the $CO_3{}^{2-}$ which is as a result of the complete decomposition of the $CaCO_3$ in the chicken eggshell [10]. IR peaks at 2654–2809 cm^{-1} and 2021–2577 cm^{-1} is assigned to the Al–Al–OH and Si–O–Al vibration of the catalyst sheet. Again well resolved strong bands at 525 and 578 cm^{-1} regions are due to Si–O stretching in the catalyst. Strong IR spectral lines characteristic for CaO are placed in the far-infrared range (~400 cm^{-1} and 290 cm^{-1}) and a weak band in the range of 500 cm^{-1} to 560 cm^{-1} indictes the formation of amorphous catalyst [6]. The BET analysis revealed a very high surface area of 340 m^2/g for the alumina supported calcium oxide catalyst with a pore volume and mean pore radius of 0.1351 cm^3g^{-1} and 7.208 Å respectively. The surface area is actually high compared to the previous studies reported in various literatures. [2] reported a surface area of 83.77 m^2g^{-1} for alumina supported calcium oxide catalyst, though the alumina loading was just 30% unlike the 100% loading used in this research. This result also varies with the report by [16] where the surface area for the calcium oxide supported with alumina in a 1:1 ratio was 82.74 m^2g^{-1}. Theodora *et al.* (2017) earlier reported that alumina has very large specific area which enhances high dispersion of active sites. According to [14], a combination of high surface area, large pore volume and smaller pore sizes allows triglycerides of different sizes to enter the pores of the catalyst, and as well gives a large surface for proper transesterification. This combination therefore, makes for a highly active catalyst. The thermal stability of the catalyst was determined via thermos-gravimetric analysis (TGA). A plot of the results obtained showed three different stages of weight loss with increasing temperature. At the first stage, the weight of the catalyst was seen to have reduced steadily between temperature ranges of 60 °C–160 °C. This was followed by a second stage between 160 °C and 360 °C which showed a drastic weight loss as the temperature was increased. According to [10], this weight loss could be attributed to loss in organic matter. The last stage of weight loss was similar to the first stage, the weight loss occurred steadily between 360 °C and 600 °C. Above 600 °C, the catalyst maintained a constant weight. This means that the catalyst is thermally stable above 600 °C.

The analysis of variance, (ANOVA) revealed that each parameter had a significant effect on the yield of biodiesel as they were all represented in the final model equation for biodiesel yield [5]. The analysis of the experimental data obtained showed that all the four factors had a considerable effect on the biodiesel yield with the impregnation ratio and the calcination temperature being of greater effect with F-value of 624.53 [10]. But in terms of interaction of the factors, the calcination temperature and calcination time had the greatest effect on the yield of biodiesel. There could be a better response in the percentage yield of biodiesel if the calcium oxide-alumina catalyst from chicken eggshell and kaolinite clay is calcined at a higher temperature and time with lower impregnation ratio and time. This result agrees with the study done by [16] where calcium oxide from chicken eggshell impregnated with gamma alumina in a 1:1 ratio and calcined at 718 °C gave the optimum biodiesel yield. Figures 1, 2, 3, 4, 5 and 6 clearly represent the effect of the interactions of the process variables used in the production of the catalyst on the yield of the biodiesel. In Fig. 1, it is observed that higher yields of biodiesel were gotten with a combination of lower impregnation ratio and higher impregnation time. From the model equation of biodiesel response, (Eq. 1) it was observed that impregnation ratio alone has a significant negative effect on the biodiesel response while the impregnation time has a significant positive effect.

$$\begin{aligned} Y = &+56.06 - 14.30 * A + 2.81 * B + 12.21 * C + 4.80 * D - 0.72 * A * B - 1.23 * A * \\ &C - 2.75 * A * D + 1.10 * B * C - 0.94 * B * D - 3.96 * C * D + 0.43 * A * B * C + 3.69 \\ &* A * C * D + 1.14 * A * C * D + 0.83 * B * C * D \end{aligned} \tag{1}$$

Since the impregnation ratio-time term of the biodiesel yield model equation is a linear term, the effect of the impregnation ratio will be more pronounced on the response. This means that loading a higher quantity of the calcium oxide on the alumina will require more time for impregnation in order to get a higher biodiesel yield. Lower biodiesel yields at lower impregnation time and high impregnation ratio could be due to improper mixing since the time of impregnation is shorter and catalyst loading higher. This result also agrees with the report by [16] that once the loading of calcium oxide on alumina increases beyond a ratio of 1:1, the yield of biodiesel begins to drop because of blockage of catalyst pore spaces by the precursor. Graphically, Fig. 2 represents the effect of the interaction of

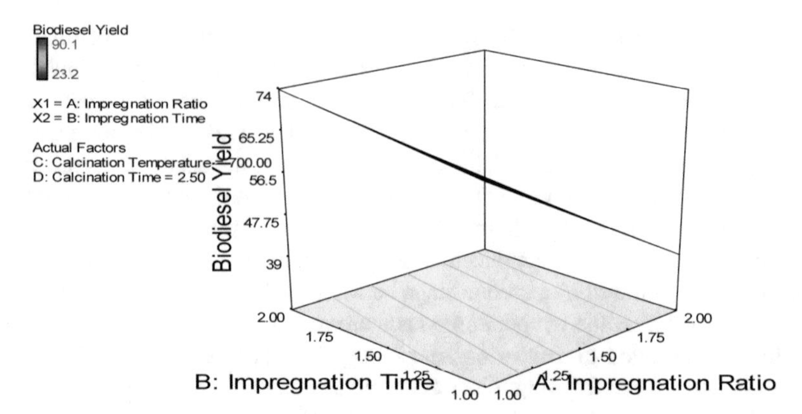

Fig. 1. A three-dimensional factorial plot on effect of impregnation ratio and time on biodiesel yield

calcination temperature and impregnation ratio on the yield of biodiesel. The biodiesel yield can be seen to have increased with an increase in temperature and a decrease in impregnation ratio. This means that in order to get a higher yield, a combination of lower ratio of impregnation ratio and a higher temperature of calcination is suitable.

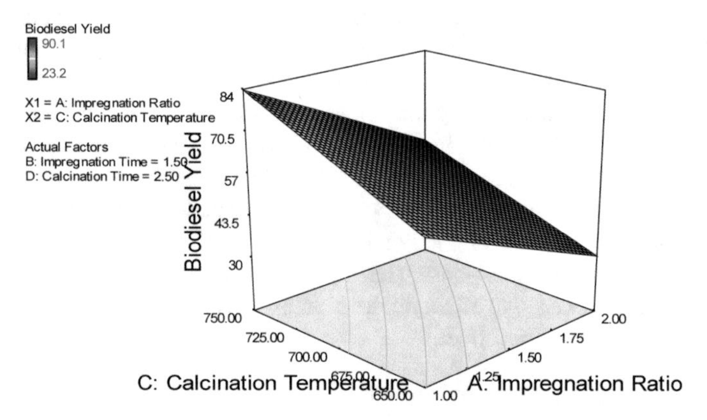

Fig. 2. A three-dimensional factorial plot on effect of impregnation ratio and calcination temperature on biodiesel yield

Figure 3 shows the effect of the interaction between the ratio of impregnation and the time of calcination. From the Figure a better biodiesel response was observed with a combination of longer hours of calcination and a lower impregnation ratio.

It was also seen that between two to three hours of calcination, the response is high with low impregnation ratio. But as the impregnation ratio increased, the response begins to drop even with long hours of calcination. This could be seen in the model equation as earlier explained. This means that the ratio used in impregnation is of utmost importance as it has a strong effect on the biodiesel response [16].

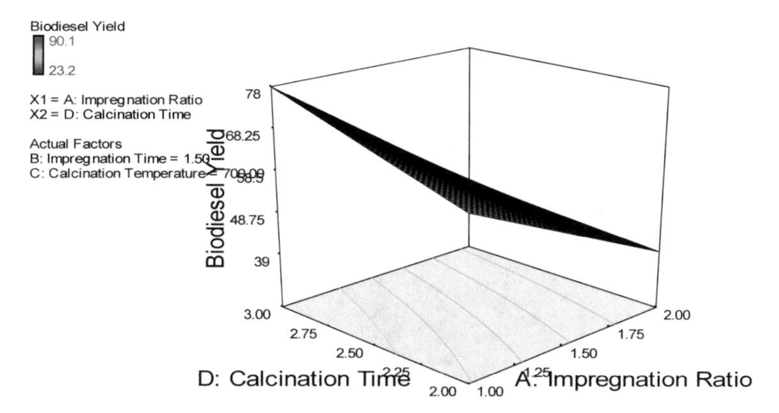

Fig. 3. A three-dimensional factorial plot for effect of impregnation ratio and calcination time on biodiesel yield

In Fig. 4, it can be seen that the yield of biodiesel increased with increase in calcination temperature and impregnation time. Worthy of mention is the fact that there seems to be a very slight difference in the yield at lower impregnation time than that at higher time.

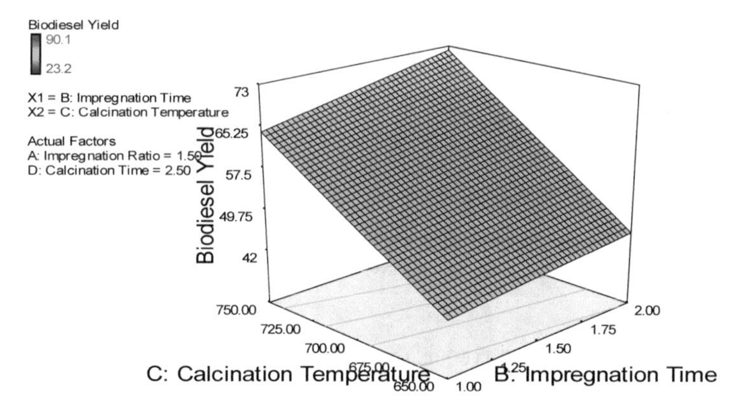

Fig. 4. A three-dimensional factorial plot for effect of impregnation time and calcination temperature on biodiesel yield

From the plot, there seems to be a greater effect on the yield from the calcination temperature than that from the hours used in impregnation. From the model equation, it can be seen that the calcination temperature-impregnation time term of the equation is linear and has a positive term. It implies that the interaction of these factors have a significant positive effect on the biodiesel response.

That is, when the catalyst is impregnated for longer hours and calcined at higher temperatures, the yield will be higher. The response from the interaction of calcination time and impregnation time is shown in Fig. 5 as positive. This can be seen as the biodiesel yield increased linearly with increase in calcination time and impregnation time.

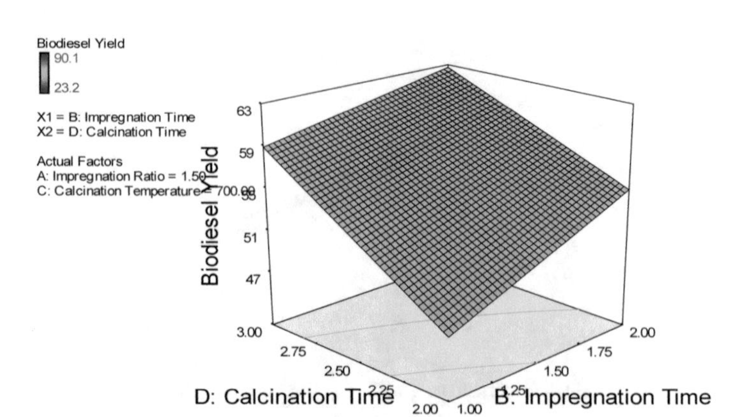

Fig. 5. A three-dimensional factorial plot for effect of impregnation time and calcination time on biodiesel yield

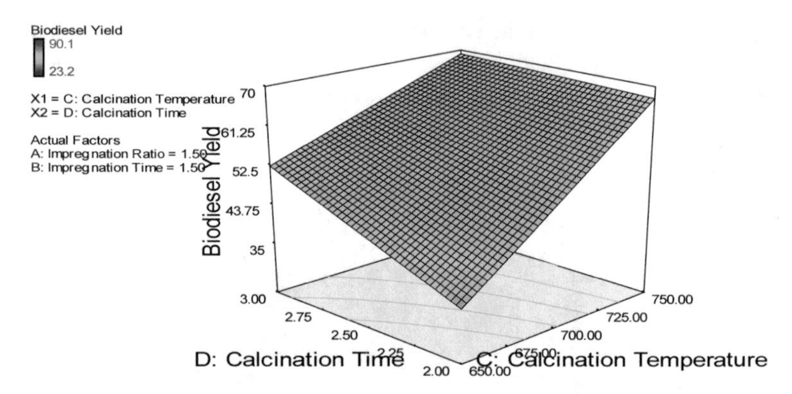

Fig. 6. A three-dimensional factorial plot for effect of calcination temperature and time on biodiesel yield

A similar occurrence is observed as the biodiesel yield increased from low to high with increase in calcination time and calcination temperature with calcination temperature giving a greater effect on the response, Fig. 6. Figure 7 showed representation of the contour plots for the biodiesel yield predicted within the ranges of the variables studied. From the plots, each contour represents two factors, at the centre points of which the biodiesel yield can be seen.

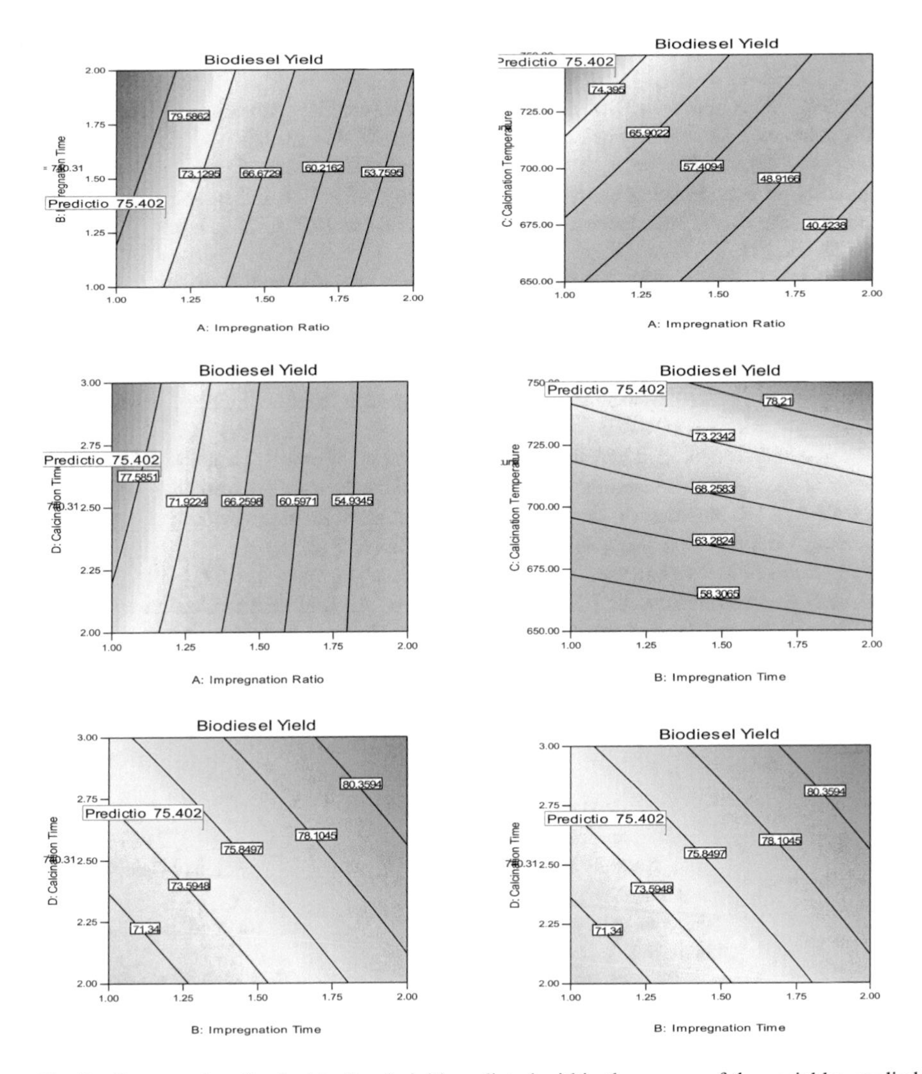

Fig. 7. Contour plots for the biodiesel yield predicted within the ranges of the variables studied

4 Conclusion

From the design of experiment using 2^4 factorial design, the interaction between the calcination temperature and calcination time had the greatest effect on the yield of biodiesel. The optimum conditions of the reaction were obtained using numerical optimization methods. These conditions are; impregnation ratio of 1.17:1, 1.32 h of impregnation, calcination at 740.31 °C for 2.62 h. The biodiesel yield from these optimum conditions was predicted to be 75.4% with 95% confidence level. The conditions for transesterification were 12:1 methanol to oil molar ratio, 6 wt% catalyst, 60 °C reaction time, and mixing speed of 250 rpm for ninety minutes.

References

1. Ajemba, O.R., Onukwuli, O.D.: Kinetic model for Ukpor clay dissolution in Hydrochloric acid solution. J. Emerg. Trends Eng. Appl. Sci. **3**(3), 448–454 (2012)
2. Asri, N., Diah, A., Puspita, S., Bambang, P., Suprapto: Utilization of waste cooking oil for biodiesel production using alumina supported base catalyst. In: 3rd International Conference on Biological, Chemical and Environmental Sciences, BCES, Kuala Lampur, Malaysia, pp. 59–63 (2015)
3. Eterigho, E.J., Farrow, T.S., Ejejigbe, S.E.: Sulphated zirconia catalyst prepared from solid sulphates by non-aqueous method. Iran. J. Energy Environ. **8**(2), 142–146 (2017). http://www.IJEE.net
4. Eterigho, E.J., Farrow, T.S., Agbajelola, D.O., Ejejigbe, S.E., Harvey, A.P.: Harnessing alternative technology for the sustainability of biodiesel production. Iran. J. Energy Environ. **7**(1), 7–11 (2016). http://www.IJEE.net
5. Eterigho, E.J., Baaki, M.A., Ejejigbe, S.E.: Optimization of process parameters for the synthesis of locally sourced alumina-supported eggshell catalyst. In: Proceedings of The World Congress on Engineering and Computer Science. Lecture Notes in Engineering and Computer Science, San Francisco, USA, 23–25 October, pp. 523–527 (2018)
6. Madejova, J.: FTIR techniques in clay mineral studies. Vib. Spectrosc. **31**, 1–10 (2003)
7. Mahesh, S., Ramanathan, A., Begum, K., Narayanan, A.: Biodiesel production from waste cooking oil using KBr impregnated CaO as catalyst. Energy Convers. Manag. **91**, 442–450 (2015)
8. Muthu, K., Viruthagiri, T.: Study on solid base calcium oxide as a heterogeneous catalyst for the production of biodiesel. J. Adv. Chem. Sci. **1**(4), 160–163 (2015)
9. Olaremu, A.: Sequencial leaching for the production of alumina from nigerian clay. Int. J. Eng. Technol. Manag. Appl. Sci. **3**(7), 103–109 (2015)
10. Pandit, P., Fulekar, M.: Eggshell waste as heterogeneous nanocatalyst for biodiesel production: optimized by response surface methodology. J. Environ. Manag. **198**, 319–329 (2017)
11. Rohazriny, R., Razi, A., Naimah, I., Nasrul, H., Abidin, C.: Characterization of Calcium oxide catalyst from eggshell waste. Adv. Environ. Biol. **8**(22), 35–38 (2014)
12. Sanjay, B.: Heterogeneous catalyst derived from natural resources for biodiesel production: a review. Res. J. Chem. Sci. **3**(6), 95–101 (2013)
13. Semwal, S., Arora, A.K., Badoni, R.P., Tuli, D.K.: Biodiesel production using heterogeneous catalysts. Bioresour. Technol. **102**, 2151–2161 (2011)
14. Sharma, Y.C., Singh, B., Korstad, J.: Application of an efficient nonconventional heterogeneous catalyst for biodiesel synthesis from pongamia pinnata oil. Energy Fuels, 1–9 (2010)
15. Sulaiman, N.B.: Response surface methodology for the optimum production of biodiesel over Cr/Ca/r-Al2O3 catalyst: catalytic performance and physicochemical studies. Renew. Energy **113**, 697–705 (2017)
16. Tan, Y., Abdullah, M., Nalasco-Hipolito, C., Taufiq-Yap, Y.: Waste ostrich- and chicken eggshells as catalyst for biodiesel production from waste cooking oil: catalyst charaterization and biodiesel yield performance. Appl. Energy **160**, 58–70 (2015)
17. Zabeti, M., Daud, W.M., Aroua, M.K.: Biodiesel production using alumina supported calcium oxide: an optimization study. Fuel Process. Technol. **91**, 243–248 (2010)

Physico-Chemical Treatment of Clinoptilolite by Chitosan for the Removal of Nitrate from Wastewater

J. Kabuba(✉)

Department of Chemical Engineering, Vaal University of Technology, Private Bag X021, Vanderbijlpark 1900, South Africa
johnka@vut.ac.za
https://www.vut.ac.za

Abstract. Physico-chemical treatment of clinoptilolite was performed by coating it with chitosan in order to use it as an adsorbent material for nitrate removal from wastewater. The characterization of the modified clinoptilolite was done by Scanning Electron Microscope (SEM), Fourier Transform Infrared spectroscopy (FTIR) and X-ray fluorescence (XRF). The results obtained showed that the removal of the nitrate from wastewater is strongly dependent on initial concentration, adsorbent dosage, pH and temperature. It was found that the initial concentration of 25 mgL^{-1}, adsorbent dosage of 4.0 g, pH 3 and temperature of 25 °C are better conditions for nitrate removal from wastewater. The equilibrium sorption isotherms were analyzed by Langmuir and Freundlich isotherms and the results indicated that Langmuir isotherm describes better the adsorption process. Thermodynamic parameters such as Gibb's free energy change (ΔG°), enthalpy change (ΔH°) and entropy change (ΔS°) were studied. The results revealed that the process was spontaneous and exothermic in nature. The pseudo-first and second order model were used in analyzing kinetic data for nitrate. The experimental data were fitted well with the pseudo second order model. It was concluded that the treatment of clinoptilolite with chitosan is a highly efficient and economic adsorbent material for the removal of nitrate from wastewater.

Keywords: Chitosan · Clinoptilolite · Nitrate · Physico-chemical treatment · Removal · Wastewater

1 Introduction

This book chapter is an extended and revised version of my published research work [1]. Nitrates (NO_3^-) are origin of nitrogen for plants growth. When nitrogen fertilizers are used to enrich the soil, nitrates can be transported by rain, irrigation and other surface water (with human and animal wastes) through the soil into ground water. Pietersen [2] reported that South African groundwater regularly has high concentration of nitrate standards, which go beyond 50 $mg{\cdot}L^{-1}$, the limit mentioned by the World Health Organization. High nitrate concentration in potable water would produce a serious menace to

S.-I. Ao et al. (Eds.): WCECS 2018, *Transactions on Engineering Technologies*, pp. 195–208, 2020.
https://doi.org/10.1007/978-981-15-6848-0_16

human health such as gastrointestinal cancer, methemoglobinemia in newborn infants and liver damage [3]. Therefore, nitrate concentration reduction in potable water to acceptable levels is mandatory. Due to the fact that nitrate is a stable and extremely soluble ion and it is strenuous to be removed by traditional processes. Different processes such as reverse osmosis, ion exchange, biological denitrification, electrochemical reduction and catalytic reduction had been developed for nitrate removal from wastewater [3]. Among these techniques, adsorption one of the frequent processes for the nitrate removal from wastewater was known as the best abundant techniques for it simple operations, easy recovery, high efficiency and cost effectiveness [4]. While numerous studies have been conducted for the nitrate concentration deduction from wastewater by adsorption techniques using varieties of adsorbents [5]. Clinoptilolite has been selected as a low-cost absorbent material for the removal of anions from wastewater. The structural framework of clinoptilolite is negatively charged for it isomorphic substitution of Al^{3+} and Si^{4+} [6, 7]. Mažeikiene et al. [4] stipulated that clinoptilolite has low nitrate removal efficiency attributable to negative charge network at its structure. Its negatively charged aluminosilicate lattice not only attracts different metal cations but it can also exhibit an affinity towards anions by suitable modifications. Treatment of natural clinoptilolite with cationic surfactant greatly modifies their surface chemistry from a net negative to positive charge [8] Surfactant modified clinoptilolite with a positive charge well attract anionic contaminant similar to nitrates by electrostatic interactions [9]. Modification of the negatively charged clinoptilolite surface can be conducted by forming layers of adsorbed cationic surfactant allows the retention of anions such as nitrate by adsorption. Scrupulous in literature, various modified techniques have been conducted to enhance the removal of nitrate anions from wastewater [4]. However, some modifiers had a potential hostile effect on environment. Chitosan is a polycationic polymer with D-glucosamine units, has received vast interest due to its particular characteristics such as low cost, good adsorption performance, biocompatibility, insolubility in water, non-toxic, biodegradability environmental friendly ascribed to the presence of the large amount of reactive hydroxyl (-OH) and amino (-NH_2) group [10]. The presence of chemical activity in chitosan can make feasible the performance of physico-chemical surfactant on clinoptilolite.

2 Materials and Methods

2.1 Preparation of Materials

Clinoptilolite and synthetic solution were prepared as described in my previous work [1]. The XRF, FTIR and SEM analysis were used to characterize the natural and modified clinoptilolite.

2.2 Experimental Procedure

Batch trials were directed to ascertain the different parameters such as pH, initial concentration, adsorbent dosage and temperature. In separately item, 10 g of modified clinoptilolite was mingled with 250 mL of synthetic solution and grasped in polyethylene flask at 90 °C for 24 h. The adsorbents were then dissociated from the synthetic solution by employing a funnel filter. 50 mL of the solution was analyzed by ion chromatography (ICS 2500) to control the nitrate concentration. All the trials were performed in triplicate and at deviation of ± 0.002. The effect of pH was at the range values of 2–9, adsorbent dosage of 0.5–5.0 g, initial concentration 20–80 mgL^{-1}. The kinetic and isotherm studies were granted at different nitrate concentrations (20–80 mg NO_3^--N L^{-1}) and (20, 30, 40, and 200 mg NO_3^--N L^{-1}), respectively. Thermodynamic studies were performed at 20, 25, 30, 40 and 50 °C. The equilibrium capacity was evaluated using Eq. (1) [11]:

$$q_e = \frac{(C_0 - C_e)}{M} xV \quad (1)$$

Where q_e (mg. g^{-1}) is the quantity of nitrate adsorbed at equilibrium. C_e (mg L^{-1}) and C_o (mg L^{-1}) are equilibrium NO_3^--N and initial concentration in aqueous solution, respectively. V (L) is the volume of synthetic solution and M (g) is the mass of coated clinoptilolite.

3 Results and Discussion

3.1 Characterization of Clinoptilolite

3.1.1 Scanning Electron Microscope (SEM)

The scanning electron microscope images of natural and modified clinoptilolites presented in Fig. 1a and Fig. 1b display the apparent casing of the clinoptilolite with chitosan, i.e., surface area is linked to the adsorption capacity of the adsorbent. The change in surface zone offers additional binding sites for the adsorbate to be adsorbed can be credited to the adsorption as a surface marvel. The microscopy surface of modified clinoptilolite as revealed in Fig. 1b has established additional pores and softened structure than the natural clinoptilolite in Fig. 1a, was for the occurrence of swelling effect. This observation approves the coating of chitosan onto the clinoptilolite surface.

3.1.2 X- Ray Fluorescence (XRF)

The elemental composition results in Table 1 revealed the constitution (%) of amorphous SiO_2 in the natural clinoptilolite was 74.0% and decreased to 65.0% after modification. The decrease in the constitution (%) of SiO_2 after modification for it dissolution possibility of amorphous SiO_2. The XRF analysis results also shows the reduction in percent composition of other elements such as Al_2O, K_2O, Fe_2O_3, CaO and MgO after modification. The presence of NH_2 was observed in modified clinoptilolite, this can be justified by the existence of amine groups in the chitosan.

3.1.3 Fourier Transform Infrared Spectroscopy (FTIR)

In Fig. 2, the bonds at the range of 4000 and 3750 cm^{-1} for the natural and modified clinoptilolite showed distinct stretching, typical of water adsorption. At the range 3500 and 2000 cm^{-1}, natural clinoptilolite showed a peak and yet again this could be because of washes out the non-clinoptilolite impurities present. The major adsorption band is

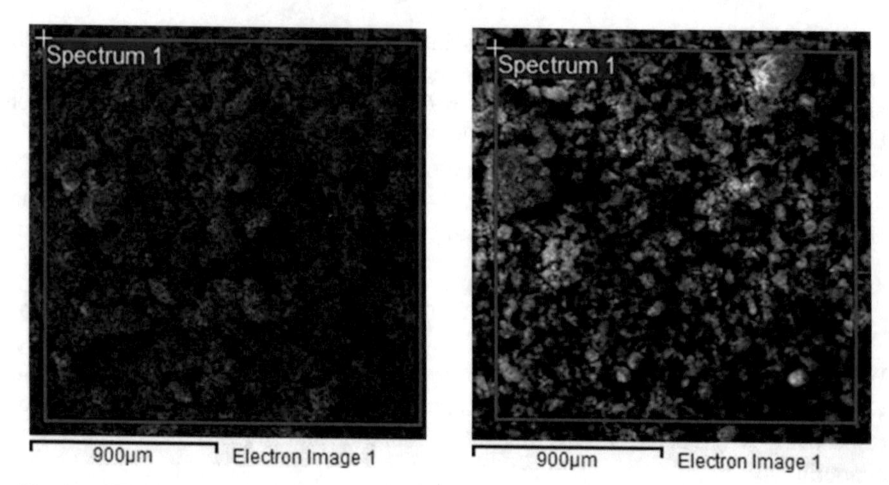

Fig. 1a. SEM images of natural clinoptilolite **Fig. 1b.** SEM image of modified clinoptilolite

Table 1. XRF analysis of the natural and modified clinoptilolite

Oxide	Natural clinoptilolite (wt%)	Modified clinoptilolite (wt%)
SiO_2	74.0	65.0
Al_2O	12.4	11.4
K_2O	3.8	2.6
Fe_2O_3	1.5	0.9
Na_2O	1.3	1.3
CaO	1.5	0.8
MgO	1.1	0.2
TiO_2	0.2	0.2
H_2O	0.7	0.7
NH_2	0.0	3.2

roughly 1979 cm^{-1} for the spectrum of pristine chitosan was attributed to intermolecular hydrogen bonds and the conjunction between O-H and N-H stretching vibration. The peaks at 1962 and 1781 cm^{-1} were attributed to C-H stretching vibration of $-CH_2$. New peak looked at 1567 cm^{-1} was assigned to the amino protonation of $-NH_2$, weak peak at 1500 indicated a double bond carbon compound. The peak intensities recorded at 1248 and 1159 cm^{-1} mentioned to $C = O$ vibration stretching of acetyl group and the stretching vibration of amine group, respectively. The peaks observed at 1055 and 1052 cm^{-1} were connected to C-O-C stretch vibration. The peaks at 1036 and 1012 cm^{-1}

linked to the stretch vibration of C-O in chitosan molecule on the position of C-3 and C-6, respectively. It was clearly seen in Fig. 3 that the peaks showed at 1000 and 960 cm^{-1} extinct for chitosan after reaction.

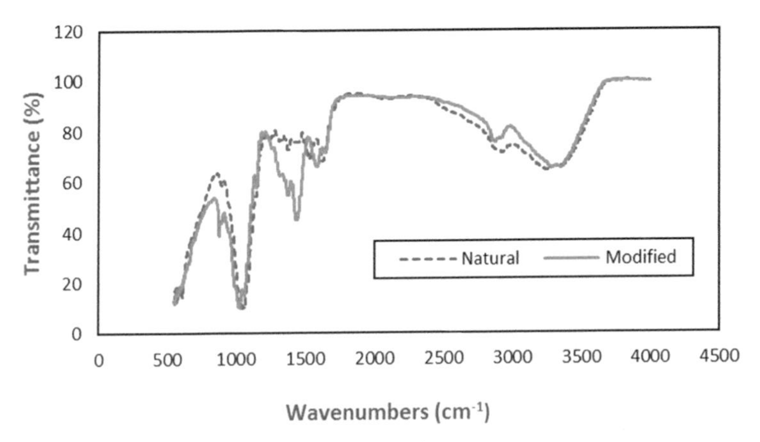

Fig. 2. FTIR spectra of clinoptilolite

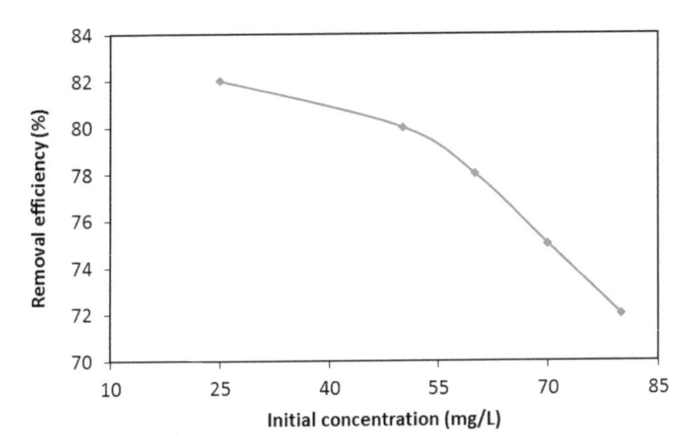

Fig. 3. Effect of initial concentration on removal of Nitrate

The intensity of the transmittance at 874 and 741 cm^{-1} in spectra, shows a strong C-O bond. The clino-chitosan spectra indicate weak peaks at 598 cm^{-1} showing a double bond of carbon compound, despite the fact that some minuscule peaks around 553 cm^{-1} shows tertiary amine bonds (CN).

3.2 Effect of Initial Nitrate Concentration and Contact Time

The effect of initial concentration for nitrate removal was determined at the following conditions: nitrate concentration from 25 to 75 mgL^{-1}, adsorbent dosage of 4.0 g and temperature of 25 °C. The nitrate removal on modified clinoptilolite decreases from 82%

to 72% with an increased in the initial concentration from 25 to 75 mgL^{-1}. For the lowest initial concentrations, maximal removal reached 82% as shown in Fig. 3. This may be ascribed to the more active sites available than the number of adsorbate species. An increase in initial nitrate concentration produces in a considerable drop in breakthrough time. At low initial concentration breakthrough occupied late and the treated volume was greater since the lower concentration gradient produced a slower transport attributable to lessen diffusion coefficient or mass transfer coefficient [9].

3.3 Effect of Adsorbent Dosage

The effect of adsorbent dosage on nitrate removal was vigorous parameter affecting adsorption capacity and effluent concentration. The effect of adsorption dosage was studied at the following conditions: nitrate initial concentration 25 mgL^{-1}, adsorbent dosage 0.5, 1.0, 2.0, 3.0, 5.0 g and temperature 25 °C. The nitrate removal percentage increased from 75% to 81% with an increased in adsorbent dosage from 0.5 to 5.0 g as depicted in Fig. 4. The increase was attributed to the accessibility of active sites and high surface area at greater dosage [12]. This increase may be due to binding of almost all nitrate into clinoptilolite. The equilibrium establishment between the ions bond to the clinoptilolite proposed that the adsorption rate of nitrate from solution augments with increasing the volume of adsorbent.

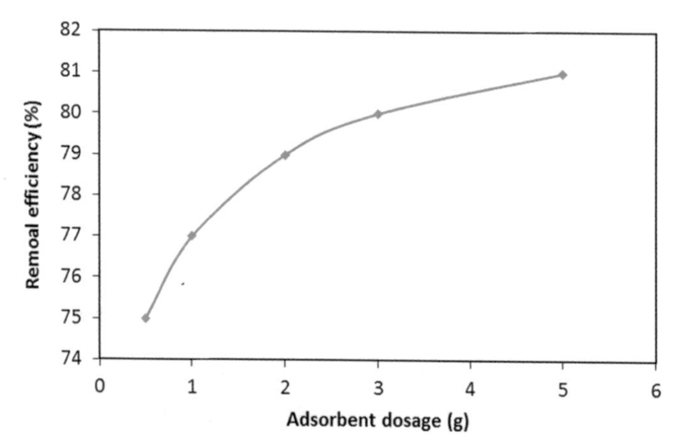

Fig. 4. Effect of adsorbent dosage

It is understandable from the Fig. 4 that the removal of nitrate augmented with increase in adsorbent mass. This is to be envisaged due to a fixed initial solute concentration, augment in total dose indicate a greater surface area and augment adsorption potential [8].

3.4 Effect of Temperature

The effect of temperature was conducted at the following conditions: temperatures 10–50 °C, nitrate concentration 25 mgL^{-1}, adsorbent dosage 4.0 g. At greater temperature, the electrostatic interaction starts to be weaker which produces the anions to become smaller promoting the adsorption of ions onto the surface of clinoptilolite with nitrate removal of 82% at 25 °C (see Fig. 5).

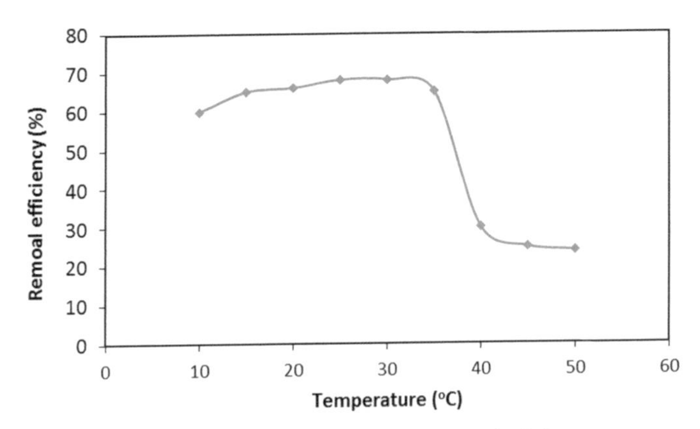

Fig. 5. Effect of temperature on removal of nitrate

Temperature decreases with removal efficiency. Temperature influences the adsorption rate by attiring the molecular interactivities and the solubility of the adsorbate. A decrease in the nitrate uptake value with the increase in temperature may be attribute to either the destruction of active binding sites of the adsorbent [11] or increasing susceptibility to desorb nitrate ions form the interface to the solution. However, augment in temperature diminished the nitrate adsorption, which might be due to an increase of the movability of the nitrate ions and a bulging effect within the inside structure of chitosan beads.

3.5 Effect of PH

The pH is apparently as a significant variable for the nitrate removal from wastewater [6]. To scrutinize the effect of pH, the experiments were ran by changing the pH values in the range of 2–9, nitrate concentration 25 mgL^{-1}, adsorbent dosage 4.0 g and temperature 25 °C. It shows in Fig. 6 that the nitrate removal decreases with an increased in pH. This implying that electrostatic attraction among having a negative charge nitrate ions and having a positive charge amine groups of chitosan were not a distinctive mechanism of nitrate removal from 2 to 9 may be attribute to the electrostatic interaction between the NH_3^+ groups of clinoptilolite and the nitrate ion. So an acidic pH value is necessary to provide -NH_3^+ groups in the clinoptilolite structure, thereby assisting the increase of nitrate adsorption. At a pH value above 8, the hydroxyl ions may engage with the nitrate anions producing in a gradual degrease in nitrate uptake [12]. The maximum nitrate removal percentage of appeared at pH 10 was attributed to the competitiveness between hydroxide and nitrate ions for adsorption sites and an enhance in diffusion resistance of nitrate case by large hydroxide ions [13]. It can be concluded that the nitrate removal could be well performed in basic condition. The maximum nitrate removal (75%) at low pH value of 2 was due to the presence of H^+ and H_3O^+ in the solution can engage with the negatively charged nitrate ions for the presence adsorption sites in chitosan which are positively charged by amine groups (NH_2 becomes NH_3^+) by electrostatic attraction [14, 15]. In acidic solution, the amine groups ($-NH_2$) on chitosan molecules are protonated and thus receive a positive charge ($-NH_3^+$), affecting the absorption capacity.

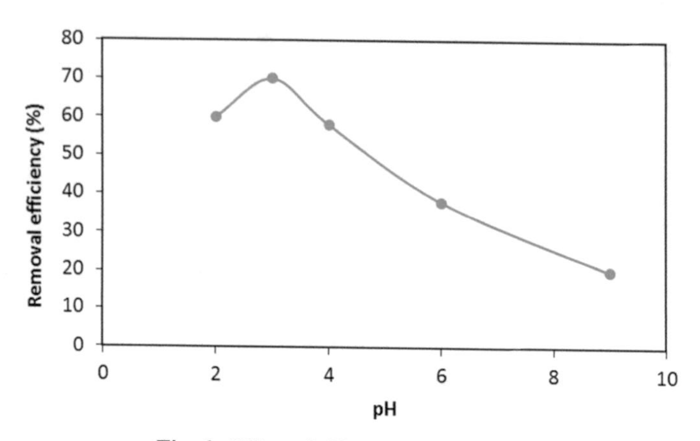

Fig. 6. Effect of pH on removal of nitrate

3.6 Adsorption Equilibrium

Adsorption isotherm is an indispensable and significant factor required to design an adsorption process. To acquire a model that will illustrate an adsorption process, it is required to institute an appropriate interconnection for the equilibrium adsorption curve. Various adsorption isotherm models have been implemented to examine experimental data and explained the equilibrium of adsorption. Equations (2) and (3) described the Langmuir and Freundlich models, respectively.

$$\frac{C_e}{q_e} = \frac{C_e}{q_m} + \frac{1}{q_m b} \tag{2}$$

$$\log q_e = \log K_F + \frac{1}{n} \log C_e \tag{3}$$

Where K_F and b are the Freundlich and Langmuir constants respectively, 1/n is the heterogeneity coefficient.

The high value of the correlation coefficient (R^2) of the adsorption isotherm values shows that adsorption of nitrate ions on modified clinoptilolite were fitted to Langmuir isotherm model as presented in Fig. 7a and Fig. 7b. This result could be due to physical as well as chemical interaction. This suggested that the adsorption of nitrate onto modified clinoptilolite was heterogeneous Freundlich compare to monolayer adsorption (Langmuir).

The equilibrium adsorption data of nitrate by chitosan reveal that the adsorption capacity increases with an enhance in equilibrium concentration and finally attains a saturated value. The adsorption isotherm model assists the comprehensive discerning of the nature of interactivity between adsorbate and adsorbent. In the Langmuir isotherm model, it is presumed that intermolecular forces lessen quickly with distance, and this conducts to monolayer coverage of the adsorbate at specific homogenous sites on the outer surface by the adsorbent. The values of the correlation coefficient stipulate that the adsorption on chitosan beads fit well to the Langmuir model.

The Freundlich isotherm model does not present fair fit to experimental equilibrium adsorption data. The parameter values of K_F and 1/n was evaluated from the intercept and slope of the plot of log q_e versus log C_e at different temperatures and are presented in Table 2. The K_F value increases with the adsorption capacity of the adsorbent. The comparatively values of the correlation coefficient (R^2) resulted from the Freundlich isotherm equation show that the adsorption process is not quite heterogenous. This stipulates that the absorption process takes place mainly by the ionic interactions between nitrate anions and amine cations despite the fact that the adsorbent surface accommodate some different groups such as hydroxyl and acetyl in the chitosan molecule [9].

Table 2. Isotherm parameters for nitrate adsorption

Langmuir isotherm		Freundlich isotherm	
q_m (mg/g)	24.620	K_F	3.7634
b (L/mg)	0.1335	1/n	0.2030
R^2	0.9811	R^2	0.8540

3.7 Kinetic Studies

The kinetic behavior of this process was conducted at pH 3 and 25 °C. This can be employed to produce the adsorption equilibrium and to establish a model that can assist in system design. Pseudo-first order and second kinetic models were produces to precisely reflect the tendency of nitrate adsorption with time and release the reaction pathway of the process. The linear pseudo-first and second order kinetic models were revealed by Eqs. (4) and (5), respectively.

$$\log(q_e - q_t) = \log(q_e) - k_1 t \tag{4}$$

Where q_e and q_t denotes the amount of nitrate ion absorbed on the modified clinoptilolite at equilibrium and time t respectively. k_1 (min^{-1}) is the rate constant of the pseudo-first order kinetics.

$$\frac{1}{q_t} = \frac{1}{K_2 q_e^2} + \frac{1}{q_e} t \tag{5}$$

Where K_2 (g mg^{-1} min^{-1}) is the rate constant. The experiment data were employed to investigate the kinetics models for nitrate adsorption onto clinoptilolite. The adsorption kinetic parameter values are presented in Fig. 8a and Fig. 8b. It was observed that the

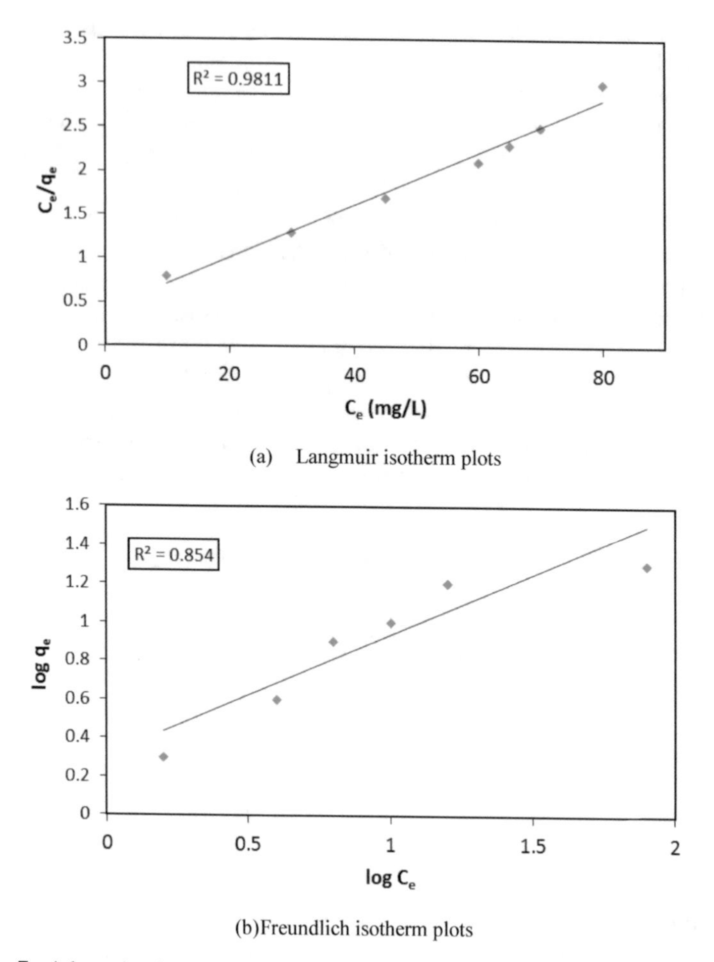

(a) Langmuir isotherm plots

(b)Freundlich isotherm plots

Fig. 7. Adsorption isotherm plots for nitrate adsorption (Langmuir and Freundlich)

higher value of R^2 (0.9897) shows the pseudo-second order model describes the nitrate kinetic study more than the pseudo-first order model with $R^2 = 0.9559$. The second pseudo-order model was more appropriate to express the kinetic adsorption behavior for nitrate into clinoptilolite.

3.8 Thermodynamic Studies

Thermodynamic factors such as Gibb's free energy change ($\Delta G°$), enthalpy change ($\Delta H°$) and entropy change ($\Delta S°$) were studied. The $\Delta G°$ (kJ mol^{-1}) value was evaluated using the Eq. (6) [16]:

$$\Delta G^{\circ} = -RT \ln K_d \quad (6)$$

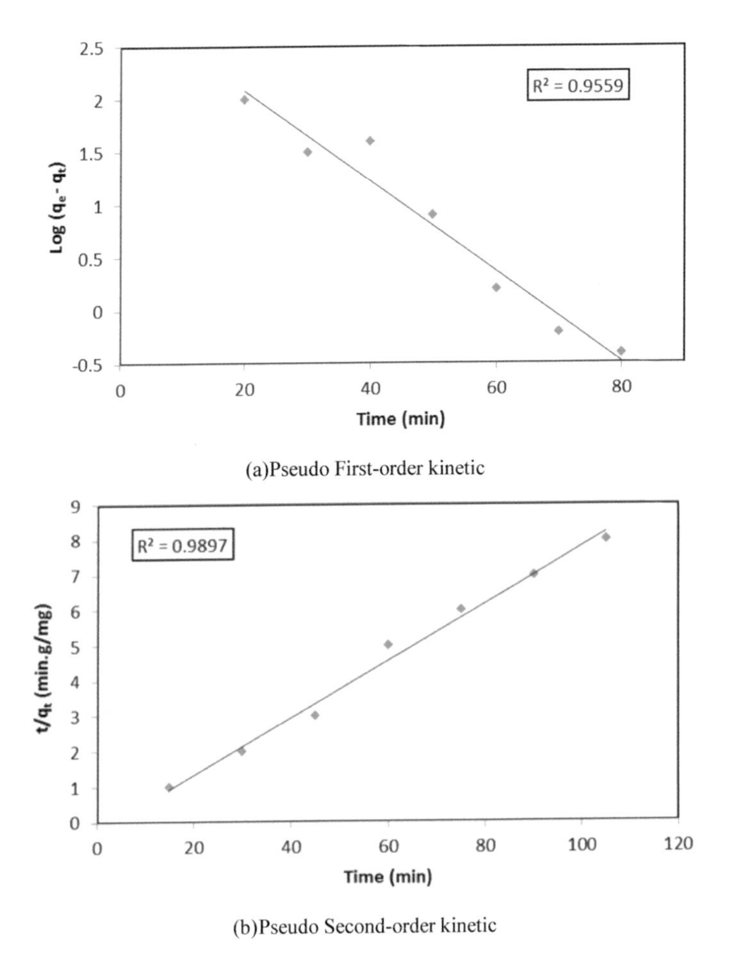

(a)Pseudo First-order kinetic

(b)Pseudo Second-order kinetic

Fig. 8. Kinetic models for nitrate removal

Where K_d (mL g^{-1}) is the distribution coefficient considering the selectivity for nitrate adsorption and was evaluated using Eq. (7):

$$K_d = \frac{q_e}{C_e} \tag{7}$$

The ΔS° (J mol^{-1} K^{-1}) and ΔH° (kJ mol^{-1}) values were evaluated from the intercept and slope of linear plot of $\ln K_d$ versus 1/T, respectively (see Fig. 9) employing the Eq. (8):

$$\ln K_d = \frac{\Delta S^{\circ}}{R} - \frac{\Delta H^{\circ}}{RT} \tag{8}$$

Table 3 presents the thermodynamic parameters for nitrate adsorption process. It was observed that the negative value of ΔH° confirms the exothermic nature of process.

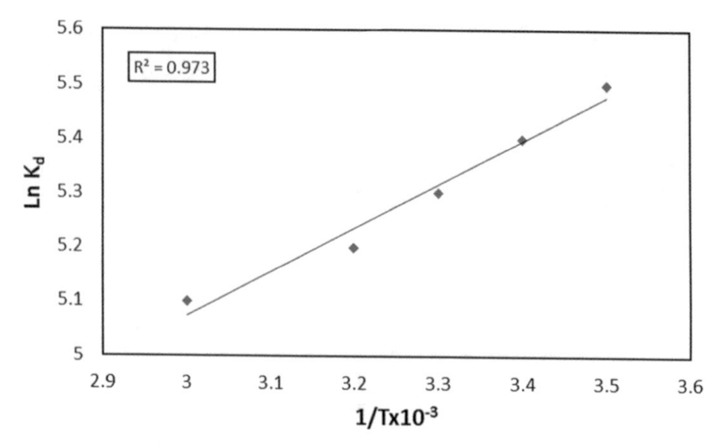

Fig. 9. ln K_d versus 1/T for nitrate adsorption

Table 3. Thermodynamic parameters for nitrate adsorption

Temperature (°C)	K_d	ΔG° (kJ/mol)	ΔH° (kJ/mol)	ΔS° (J/mol.K)
293	708.56	−11.86		33.67
298	743.60	−11.88		33.12
303	703.52	−12.25	−5.01	33.79
313	664.08	−13.50		36.71
323	658.12	−13.86		36.68

Thus, adsorption of nitrate onto surfactant modified clinoptilolite is more intensive at low temperature [17]. The negative value of ΔG° indicates that the adsorption process was spontaneous in nature and thermodynamics feasible adsorption process. The negative value of ΔG° with the higher temperature confirmed the nitrate adsorption on coated clinoptilolite was probably to happen when temperature increases. The positive value of ΔS° comprehensibility presented the affinity of nitrate ions and randomness during the adsorption process.

4 Conclusion

Clinoptilolite was successfully physico-chemical treated using chitosan to enhance his properties for the nitrate removal. The results showed that the nitrate removal efficiency decreases due to an enhance in pH while the initial concentration and adsorbent dosage increase with the nitrate removal from the wastewater. At the temperature value of 25 °C, the maximum removal was achieved at 82%. It was also found that the pH 3, initial concentration of 25 mgL^{-1} and adsorbent dosage of 4.0 g are the best removal conditions for nitrate from wastewater. Langmuir isotherm model described better fit the adsorption process. The kinetic studies were seen to obey the pseudo-second-order model. The thermodynamic investigation reveals spontaneous and exothermic nature of nitrate ion onto modified clinoptilolite.

References

1. Kabuba, J.: Modification of clinoptilolite by chitosan and application in the removal of nitrate. In: Proceedings of The World Congress on Engineering and Computer Science 2018. Lecture Notes in Engineering and Computer Science, 23–25 October 2018, San Francisco, USA, pp. 482-487 (2018)
2. Pietersen, K.: Groundwater crucial to rural development. In: Proceedings of Biennial Groundwater Conference, South Africa, 7–9 March 2005. CSIR International Convention Centre, Pretoria (2005)
3. Nujic, M., Milinkovic, D., Habuda-stanic, M.: Nitrate removal from water by ion-exchange. Croat. J. Food Sci. Technol. **9**, 182–186 (2017)
4. Mažeikienė, A.A., Valentukevičienė, M., Rimeik, A.M., Matuzevičius, A.B., Dauknys, R.: Removal of nitrates and ammonium ions from water using sorbent zeolite (clinoptilolite). J. Environ. Eng. Landscape Manage. **16**, 38–44 (2008)
5. Zhang, L., Zeng, Y., Cheng, Z.: Removal of heavy metal ions using chitosan and modified chitosan: a review. J. Mol. Liq. **214**, 175–191 (2016)
6. Abdulkareem, S.A., Muzenda, E., Afolabi, A.S., Kabuba, J.: Treatment of clinoptilolite as an adsorbent for the removal of copper ion from synthetic wastewater solution. Arab. J. Sci. Eng. **38**, 2263–2272 (2013)
7. Kabuba, J., Mulaba-Bafubiandi, A., Battle, K.: Neural network techniques for modelling of Cu (II) removal from aqueous solution by clinoptilolite. Arab. J. Sci. Eng. **39**, 6793–6803 (2014)
8. Jaina, M., Gorg, V.K., Kadirvelu, K.: Adsorption of hexavalent chromium from aqueous medium onto carboraceous adsorbents prepared from waste biomass. J. Environ. Manag. **91**, 949–957 (2010)
9. Gouran-Orimi, R., Mirzayi, B., Nematollahzadeh, A., Tardast, A.: Competitive adsorption of nitrate in fixed-bed column packed with bio-inspired polydopamine coated zeolite. J. Environ. Chem. Eng. **6**, 2232–2240 (2018)
10. Igberase, E., Osifo, P.: Equilibrium, kinetic, thermodynamic and desorption studies of cadmium and lead by polyaniline grafted cross-linked chitosan beads from aqueous solution. J. Ind. Eng. Chem. **26**, 340–347 (2015)
11. Shen, C.S., Shen, Y., Wen, Y.Z., Wang, H.Y., Liu, W.P.: Fast and Highly efficient removal of dyes under alkaline conditions using magnetic chitosan-Fe (III) hydrogel. Water Res. **45**, 5200–5210 (2011)
12. Rao, R.A.K., Rehman, F.: Adsorption studies on fruits of Gular (Ficusglomerata) removal of Cr (VI) from synthetic wastewater. J. Hazard. Mater. **181**, 405–412 (2010)
13. Chatterjee, S., Lee, D.S., Lee, M.W., Woo, S.H.: Nitrate removal from aqueous solutions by cross-linked chitosan beads conditioned with sodium bisulfate. J. Hazard. Mater. **166**, 208–513 (2009)
14. Hamed, I., Ozogul, F., Regenstein, J.M.: Industrial application of crustacean by-products (Chitin, chitosan, and chitooligosaccharides): a review. Trends Food Sci. Technol. **48**, 40–50 (2016)
15. Thakre, D., Jagtap, S., Sakhare, N., Labhsetwar, S., Meshram, S., Rayalu, S.: Chitosan based mesoporous Ti-Al binary metal oxide supported beads for defluoridation of water. Chem. Eng. J. **158**, 315–324 (2010)

16. Sari, A., Tuzen, M., Soylak, M.: Adsorption of Pb (II) and Cr (III) from aqueous solution on Celtek clay. J. Hazard. Mater. **144**, 41–46 (2007)
17. Chatterjee, S., Woo, S.H.: The removal of nitrate from aqueous solutions by chitosan hydrogel beads. J. Hazard. Mater. **164**, 1012–1018 (2009)

Aerobic Bioremediation of Fischer-Tropsch Effluent – Short Chain Alcohols and Volatile Fatty Acids

Mabatho Moreroa(✉), Diane Hildebrandt, and Tonderayi Matambo

University of South Africa, 28 Pioneer Ave, Florida Park, Roodepoort, South Africa
msmoreroa@gmail.com, {hilded,matamts}@unisa.ac.za

Abstract. Aerobic degradation was used as a treatment method to reduce the high chemical oxygen demand (COD) found in Fischer-Tropsch (FT) wastewater. The compounds investigated were short chain alcohols (SCA) and volatile fatty acids (VFAs), they contribute up to 87.4%. When released into the environment, such high strength COD can cause detrimental effect to the environment. Synthetic FT wastewater were prepared in a mineral salt solution comprising of only SCAs and VFAs and a COD of 67.9 gCOD/L. Parameters investigated were temperature and substrate concentration (COD). A gas-chromatograph revealed that VFAs were degraded faster than SCAs. Bacteria found in natural wetland situated east of Gauteng province in South Africa and FT wastewater plant, were collected, studied and used in this study. It was observed that degradation was favoured at 35 °C with 90% COD removal within three days. At substrate concentrations of 0.13, 0.73 and 1.5%, the highest COD reduction was 91, 49 and 24% respectively. The isolates were sent for 16S rRNA sequence analysis which revealed that *Bacillus* sp. was the dominant species for degrading these compounds.

Keywords: Aerobic degradation · Chemical oxygen demand · Fischer-Tropsch effluent · Short chain alcohols · Volatile fatty acids

1 Introduction

South Africa has been deemed as a water scarce country. With the country classified as semi-arid, there are limited fresh water resources, thus limiting the available water resources [1]. It has been projected that by the year 2025, most countries in the eastern and southern Africa will have limited water availability. Twelve African countries will be limited to about 1000 m^3/person/year of fresh water (thus 2.7 L/person/day), furthermore, 460 million people will be at risk of water stress [1]. In 2018, one of the biggest cities in South Africa, Cape Town, was experiencing the worst drought in 100 years, with the dams supplying the city below 25% [2]. This led to water restrictions of 50 litres per day per person, which was a further decrease from a previous 87 litres per day per person. It was reported that from May 2018, water supply in the city was to be cut off and residents would have to visit one of the 200 water collection sites for water collection [3].

S.-I. Ao et al. (Eds.): WCECS 2018, *Transactions on Engineering Technologies*, pp. 209–218, 2020.
https://doi.org/10.1007/978-981-15-6848-0_17

The National Water Act, 1998 (Act 36 of 1998) of South Africa requires a balance between using and protecting water resources to be achieved. One of the ways of achieving this, is through reclaiming and reusing treated wastewater which will in turn create an alternative water source for irrigation and other uses of water that do not require clean fresh water. This will assist in reducing the demand for portable water sources utilized for drinking. It has then become necessary to find the biggest contributors to water pollution and tackle the problem from the source. One of the biggest contributors to water pollution has been identified as the petrochemical industry. More specifically the Fischer-Tropsch (FT) technology. The FT technology is a method of converting synthesis gas containing hydrogen and carbon monoxide into hydrocarbon products as demonstrated in Eq. 1 [4–6].

$$CO + 3H_2 \leftrightarrow CH_4 + H_2O \tag{1}$$

Companies such as South African synthetic oil liquid (SASOL) and PetroSA in South Africa and Linc Energy in Australia use this technology to produce hydrocarbon products which are good for economic growth [4, 5]. However, it has been reported that for every ton of crude oil produced from the FT process, there is 1.1 to 1.3 tons of wastewater produced, which is the largest volume liquid product from the process [8–10]. Wastewater generated from this process contains high amounts of organic matter (30 gCOD/L); which comprises of alcohols (84.8%), volatile fatty acids (10.7%), hydrocarbons (4.50%) and low pH value (pH = 3.0) [7]. When discharged into the environment, this water can cause detrimental effects on the aquatic life. Short-chain volatile fatty acids i.e. acetic acid, propionic acid, butyric acid and valeric acid are of great environmental concern when discharged from industry or present in wastewater [8]. These compounds are characterised by foul odour (especially butyric acid), which is unpleasant in the environment [9]. Managing this type of wastewater has been difficult to maintain due to the toxicity and stubborn nature of the compounds [10]. Biological methods have been employed to treat this type of wastewater, but the challenge has been that the process is slow, taking between 140–180 days to degrade [11].

Biodegradation of this type of wastewater has been done using anaerobic systems, which require skilled labor to maintain, require lengthy start-up time (2–4 months) and sensitive to temperature changes [12, 13].

Due to cost implications and time constraints, it was then necessary to study the biodegradation of the compounds found in the FT wastewater in an aerobic environment. The results from this study can then be optimized and used to address the problems associated with the treatment of this wastewater in a larger scale. It has been reported that aerobic systems are used to treat wastewater with high organic loads to achieve high degree treatment efficiency and excellent effluent quality [12].

2 Experimental

2.1 Preparation of Synthetic Fischer-Tropsch Wastewater

Synthetic wastewater was prepared in a mineral solution containing: NH_4Cl (0.95 g/L), $MgCl_2.6H_2O$ (0.1 g/L), $CaCl_2.2H_2O$ (0.05 g/L), K_2HPO_4 (0.70 g/L), $NaHCO_3$ (8.4 g/L), resazurine (0.1% w/v, 1 mL/L), and metal solution (10 mL/L). The latter contained: $FeSO_4.7H2O$ (0.55 g/L), $MnSO_4.H_2O$ (0.086 g/L), $CoCl_2.2H_2O$ (0.17 g/L), $ZnSO_4.7H_2O$ (0.21 g/L), $NiCl_2.6H_2O$ (0.02 g/L), $NaMoO_4.2H_2O$ (0.01 g/L), H_3BO_3 (0.019 g/L), and nitriltriacetic acid (4.5 g/L) in distilled water [14]. Compounds found in FT wastewater as indicated in Table 1 [14], which served as carbon source (CS) were added to the mineral solution. The stock solution, with COD 67.9 g/L, was used to investigate the effect of compound concentration by serial dilutions.

Table 1. Preparation of synthetic wastewater

	Compound	Volume (mL/L)
Short chain alcohols (SCAs)	Methanol	3,01
	Ethanol	3,20
	Propanol	2,87
	Butanol	2,12
	Pentanol	1,22
Volatile fatty acids (VFAs)	Acetic acid	0,962
	Propanoic acid	0,696
	Butanoic acid	0,361
	Pentanoic acid	0,105
	Hexanoic acid	0,026

2.2 Sample Collection

Sludge was collected from a natural wetland in the East of Johannesburg, South Africa (GPS coordinates: −26.219806, 28.481750) using 5.0 L autoclaved schott bottles. The sludges were stored at 4 °C and used during this study.

2.3 Culturing Bacteria

All apparatus was autoclaved at 121 °C and for 20 min and media inoculation was carried out in a laminar flow hood. Nutrient broth and agar obtained from Sigma Aldrich were used to prepare agar plates. An inoculation loop was used to transfer the bacterial consortium from the sludges onto the agar plates using the streak plate method. The plates were incubated at 35 °C overnight for cell multiplication. Conical flasks were prepared with 50 mL mineral solution [14] and inoculum from the agar plates. The flasks were incubated at 35 °C with shaking 100 rpm for 48 h, to obtain an optical density, OD_{600} = 1.

2.4 Streak Plate Technique

Single colonies were picked from the prepared plates and transferred to Nutrient Agar plates using a sterile inoculation loop. The colony shape, elevation, opacity, pigmentation and texture of each isolates were observed. The plates were incubated at 35 °C for 24 h. The pure cultures obtained from a series of plate transfers, were used for DNA extraction and further molecular characterisation.

2.5 Effect of Substrate Concentration

Conical flasks were prepared with the stock solution and serial dilutions were done to vary the concentration of substrates (0.13, 0.73 and 1.5% v/v) in the mineral solution. An ultra-violet (UV) spectrophotometer supplied by ThermoFisher Scientific (GEN10S UV-Vis) was used to measure the OD of the samples within the media. A thermo reactor supplied by Lovibond® (RD125) was used to digest samples before reading the COD. A COD kit supplied by Lovibond® (MD 600) was used to measure the COD readings during the study. Table 2 shows the experimental set-up that was followed.

Table 2. Preparation of experimental set-up

Assay	Media (ml)	Media with CS (ml)	Inoculum (ml)
ISC	50	0	0
SC	50	0	1
IC	50	5	0
Test	50	5	1

ISC: Inoculum and substrate control, SC: Substrate control, IC: Inoculum control, Media: Mineral solution only, CS: Carbon source.

2.6 Gas-Chromatograph Characterisation

An Agilent 7890 Gas Chromatograph system equipped with a LECO Pegasus 4D Time of Flight mass spectrometer detection and a Gerstel multipurpose sampler was used for the determination of SCAs and VFAs. A Stabilwax-DA column 30 m × 0.25 mm × 0.25 μm (Restek, USA) was used for all the separation of standards and extracts. Helium was used as the carrier gas at a flow rate of 1.4 mL/min, the initial oven temperature was set 50 °C and was kept there for 0.5 min, then raised to 240 °C by 10 °C/min and held at this temperature for 2.5 min. The injected sample volume for GC analysis was 1 μL at a split of 1:10 and the run time for each analysis was 22 min. The secondary oven and modulator temperature offset were set at 10 °C. The mass spectrometry conditions were set as follows: the auxiliary temperature at the interface was 240 °C; solvent delay of 0 min; Ionization: Electron ionization at −70 eV; the source temperature was 250 °C; stored mass range: 30 to 300 μ; acquisition rate: 8 spectra/second and a detector voltage offset of plus 250 V to the optimized detector voltage.

3 Results and Discussions

3.1 Effect of Substrate Concentration

The stock solution with initial COD of 67.9 g/L was diluted using mineral solution to prepare three concentrations in triplicates i.e. 0.13, 0.73 and 1.5%. Daily measurements of OD were taken for eleven days to investigate the behaviour of cell multiplication in each flask as shown in Fig. 1.

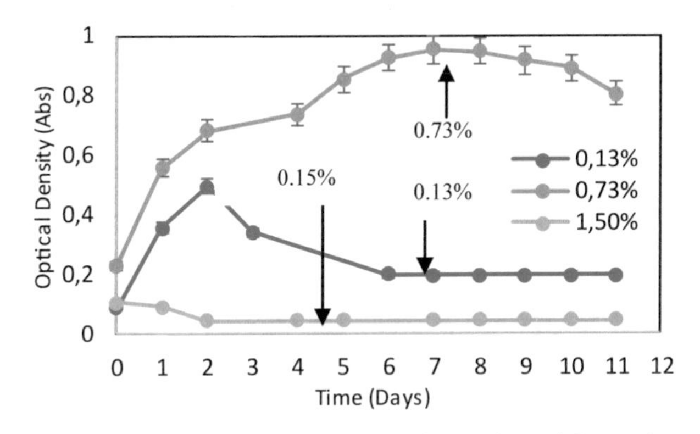

Fig. 1. Effect of substrate concentration on bacterial growth

An increase in biomass was observed in the flasks that contained 0.13 and 0.73% carbon source. The OD in the flask that contained 0.13% substrate, dropped on the third day of incubation. This was an indication that the organisms multiplied until nutrient factor limited their growth due to the depletion of nutrients [15]. When the substrate concentration was increased to 0.73%, it was observed that the OD readings continued to increase until the seventh day, then gradually started to decrease. This implied that by increasing the concentration of the substrates, the organisms were able to live longer within the media. When the substrate concentration was doubled to 1.5%, there was no growth. This was due to the environment becoming toxic, thus inhibiting the growth of bacteria.

These results were further accompanied by COD measurements, which were done to measure the amount of substrate within the solution and associate it with the behaviour of microorganisms. This is illustrated in Fig. 2.

After 3 days of incubation, 90.7% of the COD was reduced in the flask that contained 0.13% carbon source. This was followed by 49.4% and 24.98% COD reduction at substrate concentration of 0.73 and 1.5% respectively. While comparing the growth rates with COD reduction, it was noted that the organisms present in the flask that contained 0.13% substrate degraded the compounds up to day three and then stopped multiplying (indicating a death phase). This was evident by the constant COD values in Fig. 2, from day three onwards. Again, it was observed that when the concentration was increased from 0.13 to 0.73% (5.9 to 33.1 gCOD/L), the organisms continued to survive until day eight and started to die thereafter. Although the organisms were not degrading the

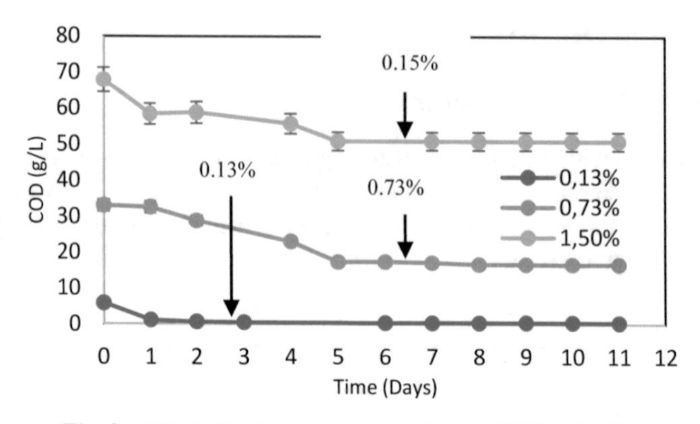

Fig. 2. Effect of substrate concentration on COD reduction

substrate as fast as the 0.13% concentration, they continued to live in the strong and toxic environment. Cells need a source of carbon for multiplication and growth. This is usually provided for by a carbon-based substrate. The substrate will provide energy to serve the purpose of cell functionality and growth. However, the amount of substrate, usually measured in concentration, needs to be monitored to ensure efficient growth of microorganisms. Too much substrate can create a toxic environment which would inhibit microbial degradation and ultimately die [15]. While too little substrate can create competition in the environment and lead to death, because of little nutrients. A study done by Hrapovic and Rowe [11] showed that the degradation of VFAs has been reported to be a slow process, taking between 140 to 180 days. This is because this contaminants are hard to manage due to their toxicity and stubborn nature [10]. From this study, we were able to find organisms that can tolerate the toxic environment of SCAs and VFAs which can be used in nature to bioremediate such pollutants at a much faster rate. The last set of experiments carried out with a concentration of 1.5% (67.9 gCOD/L) showed the least COD reduction. A maximum of 24% COD was removed throughout the study. These results validated the observation made in Fig. 1, where no growth of microorganisms was observed in these flasks. In conclusion we can deduce that the toxicity of the environment inhibited the growth of the microorganisms and the process of biodegradation.

3.2 Effect of Temperature

A study on the effect of temperature on COD reduction was done using three incubators at three different temperatures (15, 25 and 35 °C). Daily measurements of COD were done in triplicates in all the flasks as indicated in Fig. 3. It was observed that at the highest temperature (35 °C), highest COD removal (91.7%) was achieved. At lower temperatures of 25 and 15 °C, the highest achievable COD removal was 85.6 and 68% respectively. This is an indication that the biodegradation of SCA and VFA's was done by mesophylls. On one hand, extreme elevated temperatures can cause proteins found in the cell to break or denature. On the other hand, low temperatures can also have a negative effect on the cell and membrane. At low temperatures, the tails of the phospholipids in the cell membrane become more rigid and less fluid, disabling the necessary nutrients to enter the cell membrane, which also affects the growth rate of cells. During prolonged periods of colds, the fluid inside the membrane crystalizes and may pierce the cell membrane, killing the cell. The same observation was also made by Sui et al. (2016) and Kruglova et al. (2014) who reported that at low temperatures, the growth of bacteria is reduced and enzyme activity is retarded [16, 17].

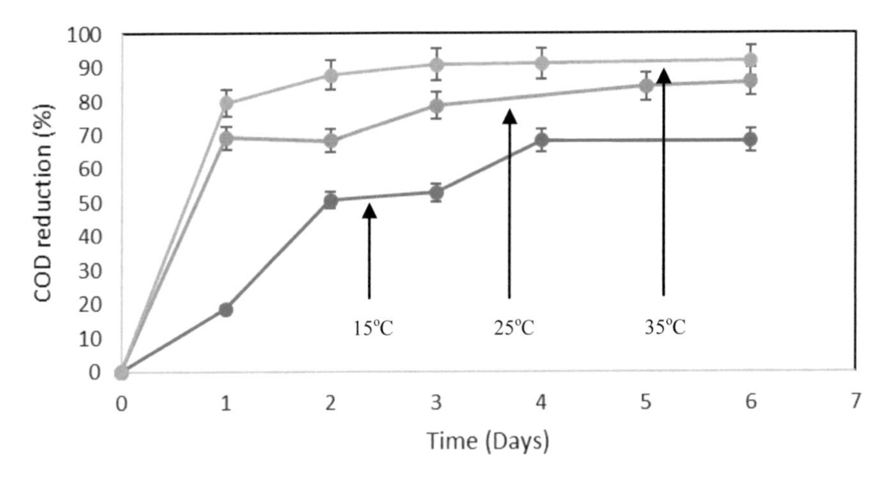

Fig. 3. Effect of temperature on COD reduction

To validate the effect of temperature on bacterial growth, four plates were prepared with nutrient agar and inoculated with bacteria, using the streak plate technique. The plates were incubated for 24 h and are shown in Fig. 4.

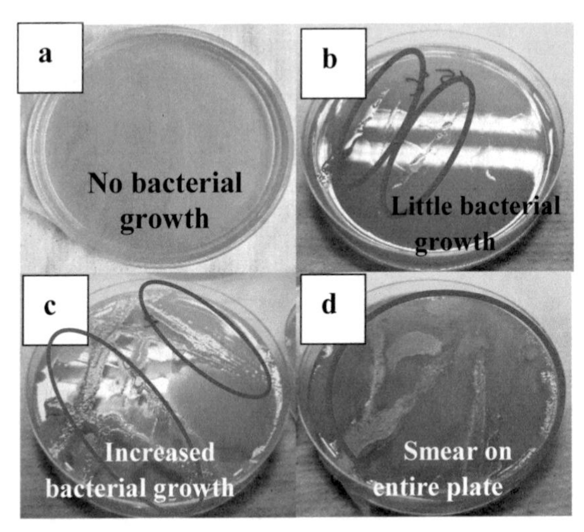

Fig. 4. Agar plates of bacteria grown at different temperatures. a) Control, b) 15 °C, c) 25 °C, d) 35 °C.

The control plate, with no inoculum, had no growth after 24 h (Fig. 4a), indicating that they were prepared in a sterile environment, thus no foreign organisms were present during preparation. Plates inoculated with bacteria showed a directly proportional relationship in growth and temperature. The growth of inoculum was identified by the smear on the plates as circled in red on Fig. 4b and Fig. 4c.

The entire plate which was incubated at 35 °C had inoculum (Fig. 4d). This validated the observation made in Fig. 3, and the conclusion that the organisms involved in this study were mesophilic. Therefore, at elevated temperatures, cells tend to function better than at low temperatures. This is because a cell membrane is formed by two layers of fatty acid membranes called phospholipids. Phospholipids within a membrane help to keep it fluid and semi-permeable so that all the necessary nutrients can enter the cell membrane, whilst harmful substances are kept out [18]. Elevated temperatures increase the fluidity of the cell membrane and thus more nutrient can penetrate the cell membrane and increase the growth rate of the cell. The opposite is then experienced at lower temperatures [18].

3.3 GCMS Characterisation

Characterisation of compounds was done daily for five days. The concentration of the compounds, which served as metabolites for the organisms was measured to investigate how each compound was being degraded from the stock solution.

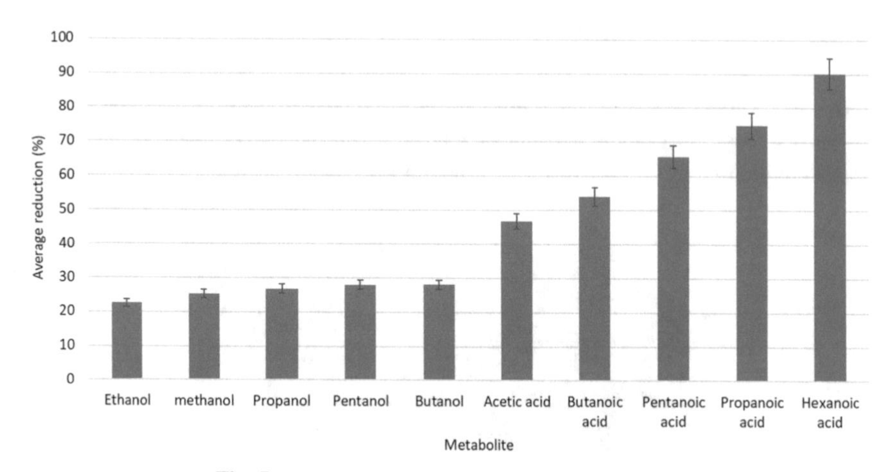

Fig. 5. Average compound removal over five days

It was observed that after five days of incubation, the compound that was degraded the least was ethanol, and the most degraded was hexanoic acid. A closer look revealed that the organisms degraded more of the VFAs than SCAs. Lindahl et al. (2017) pointed out that the degradation of VFAs is more enhanced by the addition of SCAs [19]. The study revealed that the cells that grow in a medium containing the two substrates (VFAs and alcohols) tend to have a higher specific Adenosine triphosphate (ATP) consumption rate, which is induced by acetate and proton efflux, as opposed to cells that are grown in VFAs alone. This implies that the combination of SCAs and VFAs is a good one when trying to reduce such contaminants in the environment. It was also observed from Fig. 5 that the removal efficiency of compounds increased with an increase in molecular weight. This was also observed in a recent study done by Kim et al. (2019) where an electron beam irradiation method was used to decompose the fatty acids. It was reported that the removal efficiency increased with an increase in molecular weight in the following order: acetic acid $<$ propionic acid $<$ butyric acid $<$ isovaleric acid $<$ valeric acid [9]. It cannot be ruled out that SCAs were not being degraded, as it has been reported that the higher molecular compounds are broken down into the smaller ones [9]. Thus, acetic acid breaks down into ethanol, butanoic acid into butanol, pentanoic acid into pentanol, propanoic acid into propanol and hexanoic acid into hexanol. This implies that in addition to the initial concentrations of alcohols, there is constantly an addition from the breakdown of VFAs, making it seem though as if alcohols are not being degraded. The degradation products detected by the GCMS indicated an increase in hexanol, which is the degradation product of hexanoic acid (results not shown).

3.4 Molecular Characterisation

According to biochemical characteristics and 16S rRNA sequence analysis the isolates were identified as *Lactobacillus murinus*, *Bacillus amyloliquefaciens*, *Bacillus velezensis*, *Lysinibacillus* sp., *Bacillus pumilus*, *Enterobacter xiangfangensis*, *Enterobacter hormaechei* subsp. and *Enterobacter cloacae* strain. The bacillus strain was the dominant species in the 16S rRNA identification. It has been reported that this strain is capable of growing in different types of wastewaters, regardless the conditions of the wastewater [20]. *Enterobacter* spp. are bacterial strains closely related to the family of *Enterobacteriaceae.* They are gram negative rod-shaped strains with the *Enterobacter cloacae* complex [21]. There are 23 strains of this bacteria of which 22 were isolated from humans [22]. The presence of this bacteria may arise from the fact that upstream of the wetland, sewage is discharged into the wetland.

Acknowledgment. The authors would like to thank the National Research Foundation (NRF), Department of Science and Technology (DST), SASOL, Institute for Development of Energy for African Sustainability (IDEAS) and University of South Africa (UNISA) for making this work a success.

References

1. Alex, G., Pouris, A.: A 20 year forecast of water usage in electricity generation for South Africa amidst water scarce conditions. Renew. Sustain. Energy Rev. **62**, 1106–1121 (2016)
2. Ramphele, L.: Cold front approaching Cape Town predicts good rain for Friday, 702, Cape Town, South Africa, 07 February 2018
3. Laud, G.: Cape Town's 'Day Zero' water crisis: why is Cape town running out of water? Express, Cape Town, South Africa, 05 February 2018
4. Dry, M.E., Hoogendoorn, J.C.: Technology of the Fischer-Tropsch process. Catal. Rev. **23**(1–2), 265–278 (1981)
5. Az. Cleantech: Fischer-Tropsch Catalysts, Cleantech 101 (2013). https://www.azocleantech.com/article.aspx?ArticleID=385. Accessed 20 Feb 2019

6. Zhang, M., et al.: Biological treatment of 2,4,6-trinitrotoluene (TNT) red water by immobilized anaerobic-aerobic microbial filters. Chem. Eng. J. **259**, 876–884 (2015)
7. Beccari, M., Majone, M., Dionisi, D., Donadio, A., Addario, E.N.D.: High-rate anaerobic-aerobic biological treatment of a wastewater from a Fischer-Tropsch process (2007)
8. Caunt, P., Hester, K.W.: A kinetic model for volatile fatty acid piggery wastes. Biotechnol. Bioeng. **34**, 126–130 (1989)
9. Kim, K., Seo, S.H., Park, J.-H., Kim, T.-H., Son, Y.-S., Kim, H.W.: Decomposition of volatile fatty acids using electron beam irradiation. Chem. Eng. J. **360**, 494–500 (2019)
10. Daghio, M., et al.: Electrobioremediation of oil spills. Water Res. **114**, 351–370 (2017)
11. Hrapovic, L., Rowe, R.K.: Intrinsic degradation of volatile fatty acids in laboratory-compacted clayey soil. Contam. Hydrol. **58**, 221–242 (2002)
12. Chan, Y.J., Chong, M.F., Law, C.L., Hassell, D.G.: A review on anaerobic-aerobic treatment of industrial and municipal wastewater. Chem. Eng. J. **155**(1–2), 1–18 (2009)
13. Dry, M.E.: Fischer-Tropsch reactions and the environment. Appl. Catal. A Gen. **189**(2), 185–190 (1999)
14. Majone, M., et al.: High-rate anaerobic treatment of Fischer-Tropsch wastewater in a packed-bed biofilm reactor. Water Res. **44**(9), 2745–2752 (2010)
15. Gharasoo, M., Centler, F., Van Cappellen, P., Wick, L.Y., Thullner, M.: Kinetics of substrate biodegradation under the cumulative effects of bioavailability and self-inhibition. Environ. Sci. Technol. **49**(9), 5529–5537 (2015)
16. Sui, Q., Yan, P., Cao, X., Lu, S., Zhao, W., Chen, M.: Biodegradation of beza fi brate by the activated sludge under aerobic condition: effect of initial concentration, temperature and pH. Emerg. Contam. **2**(4), 173–177 (2016)
17. Kruglova, A., Ahlgren, P., Korhonen, N., Rantanen, P., Mikola, A., Vahala, R.: Biodegradation of ibuprofen, diclofenac and carbamazepine in nitrifying activated sludge under 12 °C temperature conditions. Sci. Total Environ. **499**, 394–401 (2014)
18. Quinn, P.J.: Effects of temperature on cell membranes. Sciencing (1988). https://sciencing.com/effect-temperature-cell-membranes-5516866.html. Accessed 06 Mar 2018
19. Lindahl, L., et al.: Alcohols enhance the rate of acetic acid diffusion in S. cerevisiae: biophysical mechanisms and implications for acetic acid tolerance. Microb. Cell **5**(1), 42–55 (2017)
20. Rodriguez, O.M., Reyna, S., Lozada, J.D., Laura, S.P., Quiroz, M.A., Bandala, E.R.: Oil refinery wastewater treatment using coupled electrocoagulation and fixed film biological processes. Phys. Chem. Earth **91**, 53–60 (2016)
21. Townsend, S.M., Hurrell, E., Caubilla-Barron, J., Loc-Carrillo, C., Forsythe, S.J.: Characterization of an extended-spectrum betalactamase Enterobacter hormaechei nosocomial outbreak, and other Enterobacter hormaechei misidentified as Cronobacter (Enterobacter) sakazakii. Microbiology **154**(12), 3659–3667 (2008)
22. O'Hara, C.M., Steigerwalt, A.G., Hill, B.C., Farmer, J.J., Fanning, G.R., Brenner, D.J.: Enterobacter hormaechei, a new species of the family Enterobacteriaceae formerly known as enteric group 75. J. Clin. Microbiol. **27**(9), 2046–2049 (1989)

Author Index

S.-I. Ao et al. (Eds.): WCECS 2018, *Transactions on Engineering Technologies*, pp. 219–220, 2020.
https://doi.org/10.1007/978-981-15-6848-0

S

T

U

Y

Printed in the United States
by Baker & Taylor Publisher Services